CITIZEN RESPONSIVE GOVERNMENT

RESEARCH IN URBAN POLICY

Series Editor: Terry Nichols Clark

Recent volumes:

Volume 1:	Coping with Urban Austerity
Volume 2:	Part A: Fiscal Austerity and Urban Management
	Part B: Managing Cities
Volume 3:	Decisions on Urban Dollars
Volume 4:	Politics of Policy Innovation in Chicago
Volume 5:	Local Administration in the Policy Process: An International Perspective
Volume 6:	Constraints, Choices and Public Policies
Volume 7:	Solving Urban Problems in Urban Areas Characterized by Fragmentation & Divisiveness
Volume 8:	Citizen Responsive Government

RESEARCH IN URBAN POLICY VOLUME 8

CITIZEN RESPONSIVE GOVERNMENT

EDITED BY

KEITH HOGGART
King's College London, UK

TERRY NICHOLS CLARK
The University of Chicago, USA

2000

JAI
An Imprint of Elsevier Science

Amsterdam – New York – Oxford – Shannon – Singapore – Tokyo

ELSEVIER SCIENCE Inc.
655 Avenue of the Americas
New York, NY 10010, USA

© 2000 Elsevier Science Inc. All rights reserved.

This work is protected under copyright by Elsevier Science, and the following terms and conditions apply to its use:

Photocopying

Single photocopies of single chapters may be made for personal use as allowed by national copyright laws. Permission of the Publisher and payment of a fee is required for all other photocopying, including multiple or systematic copying, copying for advertising or promotional purposes, resale, and all forms of document delivery. Special rates are available for educational institutions that wish to make photocopies for non-profit educational classroom use.

Permissions may be sought directly from Elsevier Science Global Rights Department, PO Box 800, Oxford OX5 1DX, UK; phone: (+44) 1865 843830, fax: (+44) 1865 853333, e-mail: permissions@elsevier.co.uk. You may also contact Global Rights directly through Elsevier's home page (http://www.elsevier.nl), by selecting 'Obtaining Permissions'.

In the USA, users may clear permissions and make payments through the Copyright Clearance Center, Inc., 222 Rosewood Drive, Danvers, MA 01923, USA; phone: (978) 7508400, fax: (978) 7504744, and in the UK through the Copyright Licensing Agency Rapid Clearance Service (CLARCS), 90 Tottenham Court Road, London W1P 0LP, UK; phone: (+44) 207 631 5555; fax: (+44) 207 631 5500. Other countries may have a local reprographic rights agency for payments.

Derivative Works

Tables of contents may be reproduced for internal circulation, but permission of Elsevier Science is required for external resale or distribution of such material.

Permission of the Publisher is required for all other derivative works, including compilations and translations.

Electronic Storage or Usage

Permission of the Publisher is required to store or use electronically any material contained in this work, including any chapter or part of a chapter.

Except as outlined above, no part of this work may be reproduced, stored in a retrieval system or transmitted in any form or by any means, electronic, mechanical, photocopying, recording or otherwise, without prior written permission of the Publisher.

Address permissions requests to: Elsevier Science Global Rights Department, at the mail, fax and e-mail addresses noted above.

Notice

No responsibility is assumed by the Publisher for any injury and/or damage to persons or property as a matter of products liability, negligence or otherwise, or from any use or operation of any methods, products, instructions or ideas contained in the material herein. Because of rapid advances in the medical sciences, in particular, independent verification of diagnoses and drug dosages should be made.

First impression 2000

Library of Congress Cataloging in Publication Data
A catalog record from the Library of Congress has been applied for.

ISBN: 0-7623-0499-5

The paper used in this publication meets the requirements of ANSI/NISO Z39.48-1992 (Permanence of Paper).
Printed in The Netherlands.

CONTENTS

LIST OF FIGURES	*vii*
LIST OF TABLES	*ix*
LIST OF CONTRIBUTORS	*xiii*

1. CITY GOVERNMENTS AND THEIR CITIZENS: AN OVERVIEW OF THE BOOK
   Terry Nichols Clark & Keith Hoggart — *1*

2. THE LEGITIMACY OF LOCAL GOVERNMENT – WHAT MAKES A DIFFERENCE? EVIDENCE FROM NORWAY
   Lawrence E. Rose & Per Arnt Pettersen — *25*

3. CITIZENS, COUNCILORS AND URBAN INSTITUTIONAL REFORM: THE CASE OF THE NETHERLANDS
   Bas Denters — *67*

4. MUNICIPAL RESPONSIVENESS TO LOCAL INTEREST GROUPS: A CROSS-NATIONAL STUDY
   Keith Hoggart — *89*

5. TRANSLOCAL ORDERS AND URBAN ENVIRONMENTALISM: LESSONS FROM A GERMAN AND A UNITED STATES CITY
   Jefferey Sellers — *117*

6. THE RESPONSIVENESS OF LOCAL COUNCILORS TO CITIZEN PREFERENCES IN STUTTGART: THE CASE OF SPORTS POLICIES
   Melanie Walter — *149*

7. PLEDGES AND PERFORMANCE: AN EMPIRICAL ANALYSIS OF THE RELATIONSHIP BETWEEN MANIFESTO

COMMITMENTS AND LOCAL SERVICES
Rachel Ashworth & George Boyne *167*

8. URBAN POLICY-MAKING IN GERMANY: THE IMPACT OF PARTIES ON MUNICIPAL BUDGETS AND EMPLOYMENT
Volker Kunz *203*

9. LOCAL GOVERNMENT GROWTH AND RETRENCHMENT IN PORTUGAL: POLITICIZATION, NEO-LIBERALISM AND NEW FORMS OF LOCAL GOVERNANCE
Carlos Nunes Silva *223*

10. ENHANCING LOCAL FISCAL AUTONOMY: THE JAPANESE CASE WITH COMPARATIVE REFERENCE TO SOUTH KOREA AND THE UNITED STATES
Yoshiaki Kobayashi *243*

11. DO NEW LEADERS RISK SHORTER POLITICAL LIVES? ASSESSING THE IMPACT OF THE NEW POLITICAL CULTURE
Terry Nichols Clark *267*

THE CONTRIBUTORS 285

LIST OF FIGURES

Chapter 1

1. Party Organization Strength: National Means and Within Nation Variations Show Two Different Patterns — 6
2. Mayors' Reports of Citizen Spending Preferences — 7
3. Cross-National Differences in Citizen Versus Group Responsiveness — 8

Chapter 2

1. Voter Turnout in National and Local Elections in Norway, 1945–1999 — 26

Chapter 5

1. The Madison Urban Service Area — 127
2. The Freiburg Region — 129

Chapter 7

1. Models of Organizational Performance — 179

Chapter 8

1. Measurement Model of Party Control — 214

Chapter 10

1. Mayoral Responses to Legislature Requests in Japan — 254

Chapter 11

1. Scatterplot of Mayor's Policy Distance from National Party — 278

LIST OF TABLES

Chapter 2

1. Regression Coefficients for Perceptions Regarding Confidence in Local Politicians and their Performance — 40
2. Regression Coefficients for Perceptions Regarding the Indispensability and Performance of Local Political Parties — 42
3. Regression Coefficients for Perceptions Regarding Confidence in and Responsiveness of the Municipal Council — 44
4. Regression Coefficients for Perceptions Regarding Confidence in the Local Administration and General Performance of Local Government — 47
5. Regression of Satisfaction with Municipal Services Against Selected Explanatory Factors — 51
6. Pearson Correlation Coefficients for Measures Reflecting Legitimacy of Local Government and Satisfaction with Municipal Services with Voting in the 1995 Municipal Election — 52

Chapter 3

1. Councilor and Citizen Orientations on Local Democracy in Seven Dutch Cities — 72
2. Councilor Beliefs in Local Political Judiciousness of the Electorate in Seven Cities — 76
3. Beliefs in Local Political Judiciousness of the Electorate for Councilors of Five Major Parties — 78
4. Councilor and Citizen Support for Democratic Reforms in Seven Cities — 80
5. Correlates of Councilor Reform Support, Results of OLS Regressions — 83

Chapter 4

1. Mean Average Score for Favorable Responses to Interest Groups — 95
2. Pearson Correlations for Activity Level and Responsiveness to Interest Groups — 96

3. Municipal Responsiveness to Business Groups – Above/Below National Mean   97
4. Municipal Responsiveness to Business Groups – Pearson Correlations   98
5. Municipal Responsiveness to Low-Income Groups – Above/Below National Mean   99
6. Municipal Responsiveness to Neighborhood Groups – Above/Below National Mean   100
7. Activity and Responsiveness to Low-Income Groups   102
8. Responsiveness to Business and Low-Income Groups   105

Chapter 5

1. Indicators of Strategies Toward Environmental Amenities   125

Chapter 6

1. Traditional and Modern Sport Policies in Stuttgart   152
2. Deviation of Representative from Public Spending Preferences by Party   154
3. Links with Differential Representative Perceptions   155
4. Measures of Dyadic Responsiveness   160

Chapter 7

1. Descriptive Statistics - Proportions of Election Pledges   178
2. Descriptive Statistics: Performance Indicators 1996/97   182
3. Descriptive Statistics: Performance Indicators 1997/98   184
4. Descriptive Statistics: Change in Performance 1996/97–1997/98   186
5. Correlations Between Performance Indicators: Education 1996/97   188
6. Correlations Between Performance Indicators: Education 1997/98   188
7. Correlations Between Performance Indicators: Education Change 1996/97-1997/98   188
8. Correlations Between Performance Indicators: Social Services 1996/97   188
9. Correlations Between Performance Indicators: Social Services 1997/98   188
10. Correlations Between Performance Indicators: Social Services Change 1996/97-1997/98   189

List of Tables xi

11. Correlations Between Performance Indicators: Housing 1996/97 189
12. Correlations Between Performance Indicators: Housing 1997/98 189
13. Correlations Between Performance Indicators: Housing Change 1996/97–1997/98 189
14. Correlations Between Performance Indicators: Environmental 1996/97 190
15. Correlations Between Performance Indicators: Environmental 1997/98 190
16. Correlations Between Performance Indicators: Environmental Change 1996/97–1997/98 190
17. Correlations Between Performance Indicators: Leisure Services 1996/97, 1997/98 and Change 1996/97–1997/98 190
18. Correlations Between Performance Indicators: Financial Services 1996/97, 1997/98 and Change 1996/97–1997/98 191
19. Relationships between Explanatory Variables 192
20. Multivariate Regression Results for Performance Levels in 1996/97 and 1997/98 194
21. Multivariate Regression Results for Performance Change Between 1996/97 and 1997/98 196

Chapter 8

1. Links of Left-Right Orientation and Party Affiliation with Spending and Revenue Preferences and Perceptions of Social Problems 207
2. Spending Preferences of Politicians and Politicians' Assessments of Electoral Preferences 209
3. Spending Preferences of Politicians and Politicians' Assessments of Electoral Preferences, by Party of Politicians 210
4. Determinants of Municipal Policy Outputs 216

Chapter 9

1. Number of Presidencies (Mayors) in Câmara Municipal 227

Chapter 10

1. Political Agents that Influence Project Decisions in Municipalities in Japan, South Korea and the USA 252

Chapter 11

1. Number of Years Served by Mayors Varies Considerably Across Countries — 273
2. Sources of Variation in Mayoral Tenure: All Countries Combined, Large Model — 275
3. Sources of Variation in Mayoral Tenure: All Countries Combined, Simplified Model — 276

LIST OF CONTRIBUTORS

Rachel E. Ashworth	Public Services Research Unit, Cardiff Business School, Cardiff University Cardiff, CF10 3EU, U.K. Email: ashworthre@cardiff.ac.uk
George Boyne	Public Services Research Unit, Cardiff Business School, Cardiff University Cardiff, CF10 3EU, U.K. Email:sbsgab@cardiff.ac.uk
Terry Nichols Clark	Department of Sociology University of Chicago 1126 East 59th Street, Suite 322 Chicago, Illinois 60637, USA Email: tnclark@midway.uchicago.edu
Bas Denters	Faculty of Public Administration and Public Policy, University of Twente, P.O. Box 217, 7500 AE Enschede, The Netherlands Email: S.A.H.Denters@bsk.utwente.nl
Keith Hoggart	Department of Geography King's College London Strand, London WC2R 2LS, U.K. Email: keith.hoggart@kcl.ac.uk

LIST OF CONTRIBUTORS

Yoshiaki Kobayashi   Department of Political Science
                        Keio University
                        2-15-45, Mita, Minato-ku
                        Tokyo, 108-8345,
                        Japan

Volker Kunz   Institute of Political Science
                        Department 'Political Theory'
                        Universty of Mainz
                        Colonel-Kleinman-Weg 2
                        55099 Mainz
                        Germany

Per Arnt Pettersen   Department of Sociology and Political
                        Science
                        Norwegian University for Science and
                        Technology
                        N-7055 Dragvoll
                        Norway
                        Email: per.arnt.pettersen@sv.ntnu.no

Lawrence E. Rose   Department of Political Science,
                        University of Oslo
                        PO Box 1097
                        Blindern
                        N-0317 Oslo
                        Norway
                        Email: lawrence.rose@stv.uio.no

Jefferey M. Sellers   Department of Political Science
                        University of Southern California
                        Von KleinSmid Center 327
                        Los Angeles, CA 90089-0044, USA
                        Email: sellers@usc.edu

Carlos Nunes Silva   Centro de Estudos Geográficos
                        Universidade de Lisboa
                        Cidade Universitária
                        1699 Lisbon Codex
                        Portugal
                        Email: carlos.silva@fc.ul.pt

List of Contributors

Melanie Walter Department of Political Science
University of Stuttgart
Keplerstr. 17
70174 Stuttgart
Germany
Email: melanie.walter@po.pol.uni-stuttgart.de

1. CITY GOVERNMENTS AND THEIR CITIZENS: AN OVERVIEW OF THE BOOK

Terry Nichols Clark and Keith Hoggart

Across the world, city governments find their basic principles of operation turned on their heads as new demands emerge from citizens, higher-level authorities and non-governmental interests. Pressures on city governments are diverse and often unpredictable. City governance research seeks to inventory and assess these pressures. This is evident in the bounty of work on reactions to fiscal stress, whether explored in terms of causes of tightened city budgets (Clark & Ferguson 1983), city reaction patterns, from sluggish to hasty (e.g. Danziger, 1980), or more recently innovative responses to these forces (e.g. Clark, 1994a). Two areas of major policy change are the rise of public-private partnerships in service provision (Pierre, 1998) and a dramatic increase in the privatization of public services (e.g. Lorrain & Stoker, 1997). Much attention on the unfolding of new forms of urban governance has focused on the production of municipal services. This work has stressed improvements in the effectiveness and efficiency of governmental actions, with commentary often couched in a framework that either lauds or decries the increasing role of (private sector) producer interests in municipal policy. This increased attention to non-governmental participants is understandable, even in nations in which local government was formerly seen largely as an administrative arm of the national government.

Most obviously, this is because of growing uncertainty over the stability of local economies, alongside major adjustments in expectations and anticipations about appropriate measures for sustaining economic fortunes. Faced with fierce competitive pressure, increasingly global, large corporations have changed production practices rapidly, shifting production sites across localities, nations and even continents in a search for lower costs, improved quality and higher productivity (e.g. Scott, 1988; Dicken, 1998). Interest has risen in the last decade or so in local tools to promote local economic growth. Such interest is heightened by the impact of job shedding in formerly core manufacturing regions, alongside the rise of new industries as propulsive sectors, and growing recognition that small businesses offer potentially high rates of employment gain (e.g. Piore & Sabel, 1984; Garofoli, 1992). With national governments battling to maintain and foster national competitiveness, even localities that formerly expected national authorities to guard local economies found a growing need to emphasize strong local action (e.g. Campbell, 1990). Not only did the stature of local economic policies grow, but so did their form, as local leaders responded to service sector growth challenges, including tourism, and competitive challenges to job creation (e.g. Chandler & Lawless, 1985; Law, 1993). Faced with growing uncertainty, interest in local government-owned and controlled initiatives has risen (e.g. DeFilippis, 1999). Hardly surprising there has been a greater research focus on tightening relationships between local government, the private sector and non-governmental organizations (e.g. Harding, 1991; Sellgren, 1991). Equally important is work on reorganizing public service provision in response to new fiscal realities and changing institutional structures (e.g. Pinch, 1989).

These changes are profound. They are also often contested, by both practitioners and academic researchers. Capturing the magnitude of change, and suggesting the potential for political controversy, was the 1999 announcement that the City of Glasgow proposed to hand over all 94,000 of its remaining public housing units to a series of small, non-profit housing associations. The decision to seek tenant approval for this disbursement[1] bears testament to drastically changed attitudes. These are defining new 'realities' in Glasgow's governmental operations (Anon, 2000), for this action comes from a pioneer public housing authority, whose politics long justified the designation 'Red Clydeside', and which is still dominated by a 'traditional' left-wing council. Further illustrative of profound change are 1990s transformations in Chicago, the most studied city in the world. The city has long been famous for the manner in which its Democratic Party 'machine' gave Mayor Richard J. Daley the 'clout' to ignore citizens and pressure groups (e.g. Guterbock, 1980). Indeed Banfield (1961) and others (e.g. Fuchs, 1992; Miranda, 1994) held that 'strong party organization' helped Chicago avoid the fiscal crisis that humbled New York in 1975. This was achieved because the mayor was in a position to say no to the many 'interest groups' that pressed claims on the city. This picture contrasts sharply with New York, whose weaker mayors sought to win support by increasing spending, and thus suffered multiple fiscal crises through the twentieth century (e.g. Tabb, 1984; Shefter, 1985). When Mayor Daley II took office in Chicago in 1989, just 13 years after his father's death, he brought an entirely new style of governance to the city. The symbolic center was no longer the 'machine', but the citizen. City government shifted to measure its achievements not by power plays, but in the quality of citizen-responsive service delivery, as gauged by student test scores or turnout in park recreation programs. The Chicago case stands for legions of other cities worldwide, where powerful parties have been deeply transformed in efforts to respond more to citizens.

These examples raise another significant strand in the new realities of urban governance, which is the role of civil society in conditioning, directing and even orchestrating what local political institutions do. It is too easy to interpret shifts in municipal structures and practices as being somehow inevitable, owing to the shifts in economic circumstances that confront local governments. After all, observers ranging from popular business journalists to diverse theorists stress that economic circumstances drive most local government action, such as Peterson's (1981) market model of competition among cities for economic development or Duncan & Goodwin's (1988) articulation of how uneven class forces produce distinctive local government policies. The early literature on globalization similarly envisaged a sharpening of financial and economic constraints on localities as global transactions increased (e.g. Harding et al. 1994). But the temptation to invoke economic circumstances as the inevitably dominant force behind local government restructuring potentially blinds analysts to other powerful sources of change.

A major concern of this volume is to redress this imbalance by demonstrating the major inputs to local government from citizens and multiple interest groups including but not limited to business groups. As Barbara Anderson, Executive Director of Citizens for Limited Taxation, the lead organization behind the Massachusetts tax limitation measure Proposition 2½, expressed it: "Our fight is not mainly about money. It's about control. *They* have to learn once and for all that it's *our money*" (cited in Audit Commission, 1987, p.7). But it is up to government and party leaders to decide if they will listen to such demands. Even after two-thirds of Boston and Massachusetts citizens voted for Proposition 2½, the then Mayor of Boston, Kevin White, declared that the citizens had not understood what they had done (as discussed below). Political leaders vary considerably in how they articulate, define, and redefine what is 'politically feasible'. It would have been inconceivable for Glasgow City Council to support privatization of public housing in the 1980s when Margaret Thatcher launched the policy as an ideological centerpiece of her housing program. But as the Millennium approached, many 'realities' changed. Why is this?

'Globalization' is the buzz-word of the hour, certainly among public officials, and it is attracting enormous attention from academics (e.g. Waters, 1995; Hirst & Thompson, 1996). Early analyses saw globalization as primarily the undermining of national and local political autonomy by world capitalist markets (e.g. Friedman, 1995). But as others have looked more closely, multiple impacts of globalization have been identified. Globalization is powerful too in the near-global reach of media and news coverage (as CNN and the BBC expand), in widening concern for human rights and international humanitarian issues, in the rise of feminism, and in the wide variety of environmental, political and social movements that started in Western Europe and North America in the 1970s. A common thread in these 'political movements' has been the global pursuit of related issues that are identified as dramatic international violations of Western egalitarian and individualist values (cf. Greenpeace). In many nations these broad influences on the general public join with global communication and the diffusion of funds, staff, newsletters and workshops for activist leaders. This is occurring in a wide range of NGOs (non-governmental organizations) that are seriously involved in social and medical services in many less-developed countries. Other organizations similarly take a more openly activist position, like the Sierra Club (which is a major force across Africa and Latin America in formulating environmental policies) or Amnesty International.

One consequence of the heightened global activity of such organizations is diffusion of broad Western conceptions of egalitarianism and individual rights. Of special importance here is growing recognition of how redefinitions of these conceptions change the 'rules of the game' for public decisions in nations and localities worldwide. Villages in India and Africa are the new frontier for such organizations, since in many respects Western countries have integrated into core policies some of the conceptions and idealism that these groups have promoted since the 1970s. Official policies on issues like non-discrimination, voting rights, combating pollution, and so on, are now the firm foundation of a wide range of governmental programs and initiatives. Even if some wish to see further progress in these directions, a key point for this volume is that the cumulative effect of such changes has been to elevate the political legitimacy and power of the citizen, compared to government officials and traditional (usually hierarchically) organized groups (like most churches, unions and political parties).

Monitoring the activities of different local participants globally has become an ongoing activity of the Fiscal Austerity and Urban Innovation (FAUI) Project. Since 1982, the FAUI Project has organized international conferences, coordinated common data collection and published results of such work. This volume grew out of a series of FAUI conferences. The graphics in this

Introduction presents some results from FAUI surveys that sharpen the points just made. Look, for instance, at Figure 1.1, which shows how much political parties vary in importance in local government worldwide. Figure 1.2 portrays the range of citizen concerns for different policy areas, ranging from locations like Poland, Israel, and Germany, where mayors report that citizens prefer substantially more services, to Canada or Conservative local authorities in the UK, whose local political leaders suggest that citizens prefer far less. Figure 1.3 contrasts citizens against organized groups in their relative impact on local government. Here we find Australia on the high end, consistent with its narrow range of local functions and a strong 'public good' or 'good government' sentiment, as opposed to having a clientelist tradition. Hungary is high too, illustrating the relative lack of civic associations in the immediate post-communist period (the survey in Hungary was completed in 1991), and strong citizen empowerment concerns. (Hungary was the only East European country to ask the full set of survey questions that permit us to construct this measure).

Of course, besides cross-national differences, there are notable divergences among municipalities inside most nations. As Elazar (1970, 1986) long ago instructed us, we cannot take the political ethos that is allied to either production relations or to civil society as locally undifferentiated, neither can we assume that local political cultures represent a level playing field on which change processes interact uniformly. For instance, as the taxpayers' revolt surfaced in the USA in the 1970s, some leaders, like the then California Governor Jerry Brown, implemented the drastic cuts that voters approved in Proposition 13 with such vigor that it was easy to forget that he had opposed Proposition 13 until the hour it passed. By contrast, Boston Mayor Kevin White, confronting a similar two-thirds vote of his citizens to cut government taxes, called on the Commonwealth of Massachusetts to loan or pay for Boston programs that risked cuts. The Massachusetts legislators, broadly sharing his political culture, approved many of his requests, which led its neighbors to term the state Taxachusetts (on uneven local responses to fiscal retrenchment in these two states, see Susskind & Serio, 1983; Saltzstein, 1986).

These examples illustrate not only the importance of the dominant political ethos or political culture, but also how it acts as a context, 'framing' the actions of specific leaders by providing them with a language and with emotive markers that label their actions as 'right' or 'wrong'. If leaders seek to introduce drastic changes from the past, they risk emboldening media critics and potential competitors for office, who may invoke their cultural heritage in an effort to destroy bold incumbents who deviate from tradition. 'Framing' is a concept used in recent social movements research. Social movements are useful to consider here

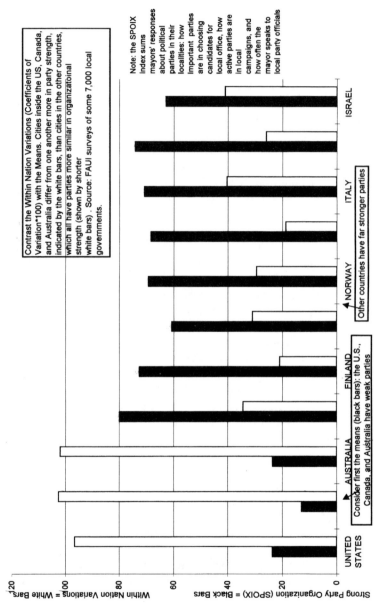

Figure 1.1: Party Organization Strength: National Means and Within Nation Variations Show Two Different Patterns.

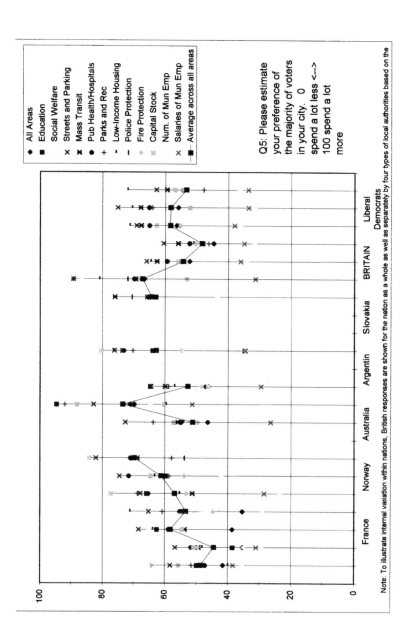

Figure 1.2: Mayor's Reports of Citizen Spending Preferences.

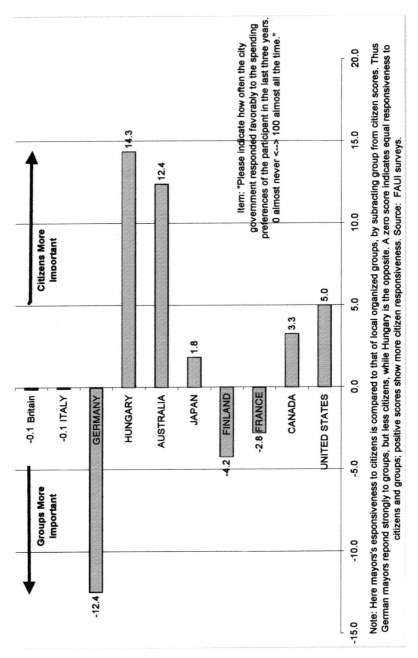

Figure 1.3: Cross-National Differences in Citizen Versus Group Responsiveness.

since they are intermediaries between citizens and more established institutions. New citizen concerns can fuel a social movement that established institutions are not addressing. Social movement theorists summarize much of their field with three core concepts that are also useful to consider for assessing citizen inputs to government (cf. McAdam et al. 1996). The first is 'framing', which we may take as the choice of language and symbolic referents, which draw on tradition and political culture. Second is the 'opportunity structure', which may broaden or narrow access by citizens or a social movement to elected officials. Opportunity structure is not static, for it may broaden in a crisis, or when a council is politically divided. The third concept is the 'degree of mobilization' of a group. In the case of citizens, this makes explicit the degree to which citizens are sharply distinct from 'pressure groups' or 'special interest groups' in more populist theories, like those of Downs (1957). Note how this contrasts with the imagery of Tocqueville or Putnam (1994, 1995) whose very label 'civic associations' implies that voluntary groups are direct extensions of concerns of the general citizenry. In some localities, as in Tocqueville's New England, or Putnam's northern Italy, citizen concerns may well be captured by civic associations, but in others, where the 'opportunity structure' is more closed, citizens may feel alienated by existing organizations, and revolt against them, sometimes fostering a new movement or organization. The polar case is the lack of civic groups independent from government in nations with authoritarian traditions (Fukuyama, 1995).

Consistent with this vision, Rider & Zajicek (1995) hold that an encouraging environment is essential for neo-liberal initiatives to work, with processes of state formation and underlying cultural values playing a key role in encouraging new governance forms. Moreover, by their very nature, innovations raise fears in people's minds, as well might be expected, given the magnitude of impacts that state policy adjustments can bring. Front and center in this regard are cuts in the labor force, which are commonly associated with privatization, contracting and lease arrangements (Adrangi et al. 1995; Patterson & Pinch, 1995). But we must resist easy assumptions about the impact of new forms of governance; for there are critical political implications behind such changes. In the case of Maharastra, for example, Chowdhury (1995) highlights how privatization measures were unable to fulfil their potential for efficiency improvements, owing to governmental attempts to maintain employment, and votes, amongst a politically active workforce (see also Evans, 1985; Chhibber, 1995). In truth, if the negative claims about new governance forms are correct, might we not expect citizen and consumer groups to express their opposition forcefully and regularly, and thus alter public policy? After all, privatization and contracting have often been accused of raising real consumer costs (Cox & Young, 1996), of disproportionately benefiting the already wealthy (Swyngedouw, 1995), and of weakening the democratic base of local government (Patterson & Pinch, 1995). Some of these charges coalesce with influential theoretical predictions, especially concerning the 'irrelevance' of citizen interests in city policy formation (e.g. Peterson, 1981).

But to restate a point already made, we risk serious error if we simply assume that political acts take place on a level playing field. This point needs emphasizing not just in geographical terms but also temporally. Michael Mann's (1986, p.7) comment that, "[t]he masses comply because they lack the collective organization to do otherwise, because they are embedded within collective and distributive power organizations controlled by others" (also Urry, 1986), might have provided an apt description of many places in the past, but how valid is it today? Even countries like the UK, which do not have a strong tradition of public protest, have seen growing citizen willingness to confront public policies they oppose since the 1980s (e.g. Doherty, 1998; Griggs et al. 1998). This may reflect less change in established institutions than a new conception of citizen concerns.

Classical rights of citizenship, as articulated by T. H. Marshall, changed drastically over the twentieth century, with the biggest changes near the century's close. Citizen concerns shifted over the century from concern with voting rights and minimal state services to an expectation that government would intervene if families mistreat their children or animals, or affect the mental well-being of a family member. Nolan (1998) has labeled this "the therapeutic state." He illustrates his point with an extreme but symptomatic example: "A family in Hawaii recovered $1,000 in compensation for the emotional distress incurred from the negligent death of their dog, Princess" (Nolan, 1998, p. 62). Changing conceptions of citizen rights suggest that public policy may have shifted less from a concern with whose 'inputs' count in policy formation (which is stressed in the 'level playing field' image), to what governments should do (is the governmental game providing a deck of cards to citizens more equally or building massive football fields replete with complex social service providers?). As citizens have grown to expect more complex and subtle social, psychological, aesthetic, and culture services from government, as well as basic economic support, 'inputs from citizens' drastically alter their meaning. They correspondingly demand closer attention from policy makers and researchers.

Shifts in service expectations have reinvigorated local government. Even in the 1960s, many northern European countries consolidated their local governments into larger metropolitan units, assuming these would provide newly expanded welfare state services more efficiently (cf. Royal Commission on Local Government in England, 1969). By the 1980s this argument was often turned on its head: the nature of welfare state services had grown so complex that they were seen by many to be better provided by smaller governments, that are closer and more responsive to citizens. 'Decentralization' became the new emblem of the day. National governments thus engaged in massive 'load shedding' of functions down to local governments, even while typically providing reduced funding. The suggestion was that local officials could raise funds to pay for these added responsibilities through imposing 'user charges' or other forms of revenue that were more citizen responsive, and by increasing the efficiency of service provision (e.g. Henney, 1984). By instituting these changes, one vision was that public service provision would operate more like private sector markets (e.g. Dunleavy, 1986; Ascher, 1987). Competition was enhanced among local governments, as well as with private firms, as seen for hospitals, schools and nursing homes. These arguments were made by policy analysts and political leaders in many countries, and of all political stripes, from Margaret Thatcher's Britain to the Communist leadership in China, to post-communist leaders in Eastern Europe (e.g. Rider & Zajicek, 1995; Swianiewicz et al. 1996; Lorrain & Stoker, 1997). In countries like Hungary and the Czech Republic, the number of local governments approximately doubled, as small local government units that Communist regimes had abolished a few decades earlier were restored (cf. Chandler & Clark, 1995; Clark, 2000). Here the prospect for more direct and powerful citizen participation in democratic governance – in smaller, local ponds – far outweighed arguments favoring potential administrative efficiency gains from larger units.

So far we have argued that the profound changes in local governance should inspire greater citizen interest and action in local government affairs. This process of provocation is being played out on a stage of citizen values which itself is undergoing major transformation. Local government leaders have seen increasing challenges to the classic municipal concerns about tax base, budget size, efficient service provision, and vote catching. Policy demands appear to rise about all of these. One reason for this is that political leaders find they are often consumed in a tide of value changes that challenge their *modus operandi*. Most evident in theoretical terms here is research on the New Political Culture (NPC) which has conceptualized these shifts in value positions (e.g. Inglehart, 1990; Clark & Rempel, 1997; Clark & Hoffman-Martinot, 1998). Central to the NPC case is the argument that citizen values are shifting away from past forms of politics, which in many countries were defined more along a left-right axis of class politics (as in much of northern Europe) or issues of patronage and favors, in a politics of clientelism (in many countries with strong agricultural traditions). Increasingly citizen values cannot be captured by the traditional value dispositions that supported class politics or clientelist regimes, with dealignment being especially notable amongst the young (also Biorcio & Mannheimer, 1995). Non-class alliances, like those based on gender, ethnicity, regionality, sexual preference, or environmentalism, provide different cultural bases for political action. Such alliances are evident in new styles of political leadership, which build on a long-held concern for fiscal conservatism but add to it social liberalism or progressivism. Such a combination was quite recently seen as hypocritical or logically impossible, but it has become a hallmark of national leaders like Blair, Clinton, and Schroeder, and thousands of mayors and council members. In Western Europe and the USA a majority of citizens similarly seem to support this New Political Culture. While new leaders might appear to introduce greater sensitivity to the shifting nature of citizen values, specific governance issues in this new highly citizen-dependent world demand closer scrutiny. New issues are raised as citizens displace other political participants. How well and how fully can leaders respond to a citizenry reflecting individualized and volatile concerns?

From the multitude of investigations on citizen values one clear, embedded message emerges: citizens are more disillusioned than enthusiastic about their governing institutions. Trust in political institutions has fallen (e.g. Ashford & Timms, 1992; Listhaug & Wiberg, 1995; contrast recent changes stressed by Fukuyama, 1995). Potentially, one explanation for this decline is that citizen values have changed so rapidly, or become so individualized, that it is increasingly difficult for political institutions to satisfy them. Alternatively, Abramson & Inglehart (1995, p. 2) caution that: "Values alone do not determine political outcomes; for they interact with economic and political forces." Put another way, values amongst the citizenry and political leaders may indeed shift in an NPC direction, but the difficulties in implementing these value changes into public policy highlight the continuing gap between citizen desires and governmental performance (Clark, 1994b). Insofar as citizens derive satisfaction or trust in government from the performance of governmental institutions, rather than the pronouncements of elected (or non-elected) officials, this disjuncture could help explain voter disillusionment. That is, citizens are no more inherently mysterious than business or political parties, but demand their increased share of attention. As new leaders work more successfully with citizens, they should see their success reflected in citizen support. A dramatic illustration in 1999 was the impeachment hearings of President Clinton by the U.S. Congress. As the weeks went by, and Congressional mobilization finally led to impeachment, the President's popularity with citizens did not decline but rose in the polls! This reflected his sharp sense of the public's specific concerns, and his ability to respond to them better than his Congressional critics (Clark, 1999).

CONCEPTUAL CONTRIBUTION OF THE BOOK

This volume is not centred around NPC ideas as such, even if these inform or underscore the chapters within it, and provide the central dilemma that underpins the volume. What the book does take from NPC research is recognition that the degree to which a New Political Culture is emerging has to be verified, with specific adoption and resistance to it charted. This requires that analysts look more closely at political systems the world over. Distinct national (and local) histories generate different political cultures, which encourage or discourage selective adoption of elements of the NPC, as well as stimulating change in public involvement for different non-governmental participants. But while culture builds on history, it is more, especially given strong globalizing pressures in recent years. Local officials worldwide have had to become more conscious of the forces and ideas that have sharpened in recent years in Western Europe and North America, particularly concerning a heightened role for citizen inputs into government processes. Related concerns of the New Political Culture like market individualism and social individualism are penetrating more deeply into previously resistant locales the world over. But specifically where, why and how is what makes these processes challenging to new leaders and controversial for interpreters.

One critical theme elaborated by many chapters is the manner in which human agency and structures combine to define 'contexts' that often redefine what is 'politically feasible' for governments to achieve. But as contexts sometimes shift rapidly with leadership changes, disentangling specifics is complex. The grounding of economic performance in political culture provides a clear illustration of this point. Thus, Contarino (1995) shows how the ease by which local political leaders could work successfully in setting policy with local unions, due to past neo-corporatist exchanges, was a key factor helping restructure the Italian textile-clothing industry to become more flexible and adopt more productive labor practices. Analogously, Heller (1995), reveals how the hegemonic position of the working class in the Indian state of Kerala provided a power base for state officials to develop corporatist relations with (local and national) capital, such that social redistribution policies accompanied economic growth promotion. What Contarino and Heller demonstrate at the local level, Alvarez and associates (1991) verify nationally. In a 1967–1984 study of advanced economies, they found that where trade unions were politically effective and had strong organization, nations governed by the political left experienced more rapid economic growth, whereas government by parties of the right required weak trade unions for stronger economic performance. This point is closely allied to the NPC idea that strong 'political' organization, in these examples exemplified by the strength (or weakness) of formal working class organizations, directly filter the impact of market relationships (and their private sector outcomes) on governmental and broader socio-economic performance (Clark & Hoffman-Martinot, 1998, p. 144, p. 133). It is not sufficient to be conscious of the dissimilar socio-political structures in which political decisions are made, nor is it adequate to assume that noteworthy distinctions between political agents provide an explanation for diverse governmental actions. These must be considered in context. Placed under dissimilar socio-political structures, similar human agents (e.g. parties of the political left), will perform differently, just as the same human agents produce uneven responses if the socio-political frameworks in which their actions take place are dissimilar (e.g. contrast the actions of right-wing parties in two contexts, those with a tradition of strong or weak party organization). The context includes socio-economic factors like wealth and occupational structure, but these take on specific meaning for governments only when they are filtered through socio-political structures and agents (Dye & Robey, 1980; Hoggart, 1987).

This volume contributes to a sharper analysis of citizens and local governments by identifying specific forms of interaction between (national and local) socio-political structures and local political agents. That is, how and when do local governing participants successfully respond to local citizens? Core issues include self-conceptions of politicians about their role in the local policy-making environment, which condition the manner and magnitude of their responses to, and even recognition of, citizen interests. In some cases politicians are little aware of citizen values (Chapter Three), in others they respond without direct pressure being placed on them (Chapter Nine), with comparison between chapters in this volume revealing significant structural differences that help explain dissimilar responses (e.g. compare Chapter Nine with Chapter Ten). Certainly there is an element of tension between what politicians want and what citizens seek (Chapter Three, Chapter Six), with leaders revealing in some contexts that differences in their vision of political desirability and practicality are critical to understanding public policy (Chapter Seven, Chapter Eight). Yet what this book seeks to draw out, through cross-national studies that directly focus on the point (Chapter Four, Chapter Five), as well as national case studies that highlight divergence in citizen and politician views (Chapter Two, Chapter Three, Chapter Six), is that the structural contextual circumstances in which politicians act critically affect their values and practices. This contextual impact extends to the very legitimacy of local government (Chapter Two, Chapter Ten), as well as to the continuance in office of leaders themselves (Chapter Eleven). Where this volume differs from most past work is in locating, measuring, and interpreting some deeper sources of political feasibility and resistance to citizen inputs by comparing cities that differ in illuminating ways.

We feature particular cases to make more general points for readers worldwide. Success as well as failure hopefully instruct us about why some policies work and warn us about why others fail. For this volume, structural distinctions across nations and municipalities are analyzed not just as socio-economic and politico-legal structures, but also in dominant socio-political practices. These provide a conceptualization for concrete, everyday political practices, which embody differentially the preferences and values of specific political agents. It is the interaction of context, preference, and actual policies that permits understanding political success or failure. Specifics of one case should not be left to stand in isolation; if we can capture the critical context, more general lessons can emerge.

The analytical setting for this volume includes the dual realities of new forms of urban governance and changing citizen values. Citizens and many organized groups question long-established patterns of political organization and practice (like left-right policy definitions and strong party leadership), while heightening political activism around single issues (the environment, sports, etc.) The end-product is less predictable local (and national) government, with more uncertain types of political coalition formation, and policy outcomes that incorporate a wider array of interests. Often relationships between citizens and their political representatives have become more taut, with declining trust in political institutions. Yet this is not inevitable. Putnam (1994, 1995), amongst others, shows how local democracy can be invigorated by incorporating active citizens into critical roles within local politics. That such occasional success is juxtaposed against a large literature criticizing new urban governance for enhancing the power of local capital, to the detriment of citizen-centred democratic values (e.g. Patterson & Pinch, 1995; Perry, 1995; Bassett, 1996), offers a deep contradiction demanding interpretation, and a key rationale for this book. Over the last decade or so, there has been a burgeoning literature on evolution in forms of city governance. It has stressed the central rationalities of local government action, dwelling on changes in policy direction, and on how key 'interests' determining policy have shifted (Harvey, 1989; Smith, 1995; Hall & Hubbard, 1998). At the same time changing citizen values have captured much attention, encouraged by some impressive research contributions (e.g. Abramson & Inglehart, 1995; Kaase & Newton, 1995). But the mainstream empirical work on political values has long stressed personal over social contextual characteristics. Our view is that these two literatures remain far too 'independent', with insufficient attention to how linkages are changing: among political structures and practices, in trends in citizen values and in changes in public policy (albeit efforts in this direction should be acknowledged; e.g. Klingemann & Fuchs, 1995; Clark & Hoffman-Martinot, 1998).[2] This volume seeks to help fill this diluted space, by presenting empirical papers that explore citizen – representative linkages in a conceptual framework that highlights context, by incorporating socio-political structures and public policy impacts.

OVERVIEW OF THE BOOK

The chapters that make up this volume present a collage of nation-specific studies and cross-national comparative investigations. Some chapters offer empirical analyses of rarely explored issues, such that their messages merit exploration in other settings. Other chapters confront more established research questions, in that the issues raised have been frequently discussed. Their distinction arises from the particular setting of the investigation, as the context is often shown to provide unexpected conclusions. These need to be fed into the general thinking and the broader literature on local government.

In terms of the scope of material covered in the book, the starting position centers on citizen views about the worthiness of local political institutions. Both Larry Rose and Per Arnt Pettersen's paper on Norway (Chapter Two) and Bas Denters' chapter on the Netherlands (Chapter Three) address growing fears about the value citizens placed on local government. In both countries, the 1990s saw worries about falling participation rates in local government elections, with corresponding fears about the declining political legitimacy of local government. Both chapters show how citizen views are more complex. In each country, certain well-grounded views amongst politicians are not accepted by the general populace. In the Netherlands, for example, citizens see little difference between the main political parties, although politicians are convinced of their critical ideological distinctiveness. Yet such divergences of interpretation and understanding do not confirm fears that local government is in crisis. In Norway, when politicians focus on electoral turnout, they may be reading the wrong game plan. Thus, Rose and Pettersen find that approval of local government service delivery plays a large part in citizen satisfaction with local government as a whole. As our commentary above suggested, citizen values are shifting in their expectations about what government provides; this becomes manifest in Norway in what legitimates local government. Legitimacy here owes much to an underlying vision of the citizen as a consumer. In the Netherlands, by contrast, legitimacy is complicated by citizens placing little value on local government, with voting in local elections couched in terms of criteria that have little local relevance. Yet efforts to reinvigorate public interest are confounded by municipal leaders not recognizing the scale of the problem, and resisting the kind of democratic initiatives that citizens favor. Similar problems of national domination of policy debates, heightened by strong national parties and their programs, have been criticized as undermining local democracy and citizen concerns in many countries with strong national parties, especially the UK, Germany and Denmark (e.g. Saiz & Geser, 1999).

Differential responses to citizen demands are taken up in the next three chapters. The first two build on cross-national comparisons. Hoggart (Chapter Four) examines disparities in the effectiveness of local interest groups in seven countries, exploring the relative responsiveness of municipalities to 'advantaged' and 'less advantaged' groups. This chapter shows that many past results from national studies of interest groups weaken in cross-national comparison. Political and socio-economic attributes conceptualized as linked to interest group effectiveness emerge as less convincing explanations of cross-national differences. This chapter uses richer data on local interest groups in cross-national perspective (from the FAUI Project) than available to past studies. One powerful general pattern emerges that is often unrecognized, locally or nationally, which is that the overall power of all interest groups varies considerably across localities. This aggregate pattern (the average amount of city impact on government of all interest groups) defines a more 'open' or 'closed opportunity structure' to an individual organized group. The more active interest groups are in general, the more local leaders believe that varied, specific interest groups are effective in getting their representations heard. It seems that a general 'atmosphere' develops in some localities that is conducive to increased responsiveness. Differences in responsiveness across nations are clearly linked to the strength of political party organization, with nations that have weaker party systems, like Australia and the USA (see Figure 1.1), seeing stronger translations of class and other non-party social divisions into municipal responsiveness. By contrast, where national party organization is stronger, the strength of local parties – not their ideological orientation – impacts most notably on interest groups getting government to respond to their preferences.

Sellers (Chapter Five) complements the general insights from Chapter Four by adding finer detail about how specific contexts shift interest group performance. He explores environmental group representation in Madison (Wisconsin) and Freiburg (Germany). Of particular relevance is the way interests are constrained by a strong local bureaucracy, upholding plans against non-governmental pressures in the German town. By contrast, the more privatized ethos that dominates the U.S. municipal system, alongside weaker translocal systems of bureaucratic, political party and interest group representation, lead to a weaker enactment of Progressive politics in Madison. Compared with Freiburg, the Madison tone, despite the city's highly progressive image in the USA, is one of heightened inequity rather than enhanced public goods provision. The responsiveness of political leaders in Madison to the values expressed by the local citizenry appears to be manifest in a 'distorted' form. Yet any 'distortion' can be traced back to systems of political organization and prevailing political values, amongst the citizenry in general as well as local leaders. As Wildavsky put it, we all act out our "cultural biases" in everything we do. In this sense each cultural view (e.g. more hierarchical or more egalitarian) has its own conception of what (for instance) is a 'level playing field' (cf. Thompson et al. 1990).

This point is carried forward by Walter (Chapter Six), who explores links between citizen values and politician preferences about sports policies in Stuttgart (Germany). Walter identifies close linkages between the two, and concludes that the political system represents citizen concerns in a significant way (see also Silva, Chapter Nine). This is not to say that the match is perfect, for Walter identifies some dissonance between politician and citizen support for 'traditional' sport policies (politicians favor more group sports like football while citizens also support individualized activities like jogging). This chapter interestingly parallels Denters' investigation (Chapter Three), as he too found politicians less keen to consider new forms of political representation.

Understanding linkages between citizens and the general public clearly requires that attention be given to the role of leaders, as this relationship is symbiotic. Focusing on one element of this relationship, Boyne and Ashworth (Chapter Seven) examine local electoral manifestos in Wales. This is one of the first studies ever to explore how local manifesto pledges (or as Americans say, 'party programs') affect changes in public service provision. Adopting a simple model to explore how policies change in the immediate years after an election outcome, Boyne and Ashworth find little support for the idea that newly elected councils institute the magnitude of changes they suggest will be forthcoming once they gain (or re-gain) office. This once more raises the question of how far citizens can trust their elected representatives. As with Denters (Chapter Three), the question of how far politicians see themselves as direct representatives of their constituents' interests, or instead consider themselves as trustees, who are given the 'right' by the citizenry to make decisions on their own judgement, lurks within these chapters.

One factor potentially weakening links between citizen preferences and urban policy decisions is the political disposition of elected representatives. This is not simply a matter of their own views, but also, as Walter (Chapter Six) shows, of how far local politicians are prepared to represent citizen views and how accurately they perceive preferences amongst the general public (or their own party's supporters). But irrespective of such considerations, research on both Germany (Chapter Eight) and Portugal (Chapter Nine) shows that ideological differences can have a major policy impact. As Kunz argues from the German case, this is an important conclusion, since it contradicts a widespread view in the German academic community that political party ideology does not intervene between institutional-legal frameworks and local policy decisions. Quite to the contrary, Kunz reports, such political impositions are more important for expenditure and employment policies than local socio-economic contexts. Given the strength of local party organizations in Germany (Figure 1.1), this pattern confirms the NPC idea that weak party organization allows for a stronger transposition of socio-economic inequalities onto municipal policies, whereas strong party organization diminishes such effects (Clark & Hoffman-Martinot, 1998, p. 133, p. 144).

Silva's Portuguese research extends this point (Chapter Nine). While noting disparities in policy enactment between locations differing in political party and socio-economic orientations, he highlights the manner in which local governments respond to increasing demands for improved public services. In a context where local government growth is the order of the day, Silva shows that local leaders respond in much the same way as in nations experiencing retrenchment and a clawing-back of state actions. Privatization, contracting out and decentralization are adopted with alacrity in order to squeeze more services out of restricted local budgets. Most certainly there are calls (and positive responses) for more funding of local services, but these are not seen to have the potential to meet faster-growing local requirements; hence the willingness to adopt innovative delivery mechanisms. Significantly, despite modest growth of interest group activity, this stance is adopted in a context of relatively limited direct citizen action at the local government level.

Further highlighting the power of cross-national divergence, Kobayashi examines how a similar picture of relative autonomy for local leaders has generated inefficient and wasteful expenditure by Japanese municipalities (Chapter Ten). With comparatively strong party organizations behind them, Japanese local leaders act with scant regard to the interests of local citizens. Municipal leaders have been drawn into this globally unusual context by a combination of legal frameworks insulating officials from effective electoral competition, plus a grant structure that essentially enables local politicians to pass municipal costs on to the national government. In this regard Japan contrasts sharply with Portugal, where new democratic practices have encouraged national government interest in local affairs, and local revenues are still drawn significantly from the purses of the local electorate. By their contrasting results, the distinct contexts for local policy in Portugal and Japan illustrate the general point that socio-political contexts clearly shift the actions of political participants.

This point is extended in Chapter Eleven, where Clark extends work on the New Political Culture by assessing its contextual impact on the years of tenure of local leaders. He documents considerable variability across nations and localities in the tenure of leaders in office. Many observers have claimed that the heightened, conflicting, and volatile demands of an NPC context of governing should reduce years in office for elected officials. The counter-intuitive finding, however, is that tenure is not reduced by an NPC national or local context. This demonstrates the considerable versatility of local leaders, many of whom have successfully adapted to the often great and conflicting demands from citizens and organized groups of an NPC context. Indeed so many have succeeded that they generally hold office as long as leaders in more ostensibly stable and traditional contexts. Whether or not socio-political environments generate new political demands and increase political conflict, the more successful city leaders often respond by adopting policy strategies and interaction modes with citizens and interest groups that enable them to continue in office. As Kunz also makes clear (Chapter Eight), both structural and human agency considerations impact on city leader actions (Giddens, 1984). The often-noted success of President Clinton in adapting to drastic crisis and change is thus shared by many other leaders in similar political contexts.

In proposing alternative models to implement for improving accountability of municipal government in Japan, Yoshiaki Kobayashi's chapter returns directly to the theme at the start of the book; namely, how the democratic accountability of local government can be improved. Across several chapters, by examining and contrasting key elements of national and local contexts, this volume provides many specific lessons. From the specific, contrasting results emerge insights – theoretical, methodological and empirical – about how citizen values may drive policy, or be ignored, in a time of turbulence and rapid cultural change for government policy making. The volume thus extends our knowledge base by conceptualizing, investigating, and interpreting specific components of the policy responsiveness of local government to its citizenry.

NOTES

1. It is a national government requirement that tenant approval is gained before such transfers of public housing can take place (Mullins et al. 1993).

2. This contextual focus was central in more general theorizing from Marx or Max Weber, and in the first voting studies by Paul Lazarsfeld and his Columbia colleagues, especially Seymour Lipset and James Coleman in the 1940s and 1950s. But later voting work more often followed the Michigan social-psychological tradition of Philip Converse and others.

REFERENCES

Abramson, P. R., & Inglehart, R. (1995). *Value Change in Global Perspective*. Ann Arbor: University of Michigan Press.

Adrangi, B., Chow, G., & Raffiee, K. (1995). Analysis of the Deregulation of the U.S. Trucking Industry: A Profit Function Approach. *Journal of Transport Economics and Policy 29*, 233–246.

Alvarez, R. M., Garrett, G., & Lange, P. (1991). Government Partisanship, Labor Organization and Macroeconomic Performance. *American Political Science Review 85*, 539–556.

Anon (2000). Revolution on the Clyde.' *The Economist* 15–21 April, 33.

Ascher, K. (1987). *The Politics of Privatization*. Basingstoke: Macmillan.

Ashford, S., & Timms, N. (1992). *What Europe Thinks: A Study of Western European Values*. Aldershot: Dartmouth.

Audit Commission (1987). *The Management of London's Authorities*. London: HMSO.

Banfield, E. C. (1961). *Political Influence*. New York: Free Press.

Bassett, K. (1996). Partnerships, Business Elites and Urban Politics: New Forms of Governance in an English City? *Urban Studies 33*, 539–555.

Biorcio, R., & Mannheimer, R. (1995). Relationship Between Citizens and Political Parties. In: H-D. Klingemann & D. Fuchs (Eds.), *Citizens and the State*, (206–226). Oxford: Oxford University Press.

Campbell, M. (Ed.). (1990). *Local Economic Policy*. London: Cassell.

Chandler, J. A., & Clark, T. N. (1995). Local Government (Around the World). In: S. M. Lipset (Ed.), *The Encyclopedia of Democracy: Volume Three*, (767–773). Washington DC: Congressional Quarterly Books.

Chandler, J. A., & Lawless, P. (1985). *Local Authorities and the Creation of Employment*. Aldershot: Gower.

Chhibber, P. (1995). Political Parties, Electoral Competition, Government Expenditures and Economic Reform in India. *Journal of Development Studies 32*, 74–96.

Chowdhury, S. R. (1995). Political Economy of India's Textile Industry: The Case of Maharastra 1984–89. *Pacific Affairs 68*, 231–250.

Clark, T. N. (Ed.). (1994a). *Urban Innovation*. Thousand Oaks, California: Sage.

Clark, T. N. (1994b). Race and Class Versus the New Political Culture. In: T. N. Clark (Ed.), *Urban Innovation*, 21–78. Thousand Oaks, California: Sage.

Clark, T. N. (1999). The Clinton Paradox: A WordWide Perspective, *Frontier*, April/May 1999: 17.

Clark, T. N. (2000). Old and New Paradigms for Urban Research. *Urban Affairs Review* (in press).

Clark, T. N., & Ferguson, L. C. (1983), *City Money: Political Processes, Fiscal Strain and Retrenchment*. New York: Columbia University Press.

Clark, T. N., & Hoffman-Martinot, V. (Eds.), (1998). *The New Political Culture*. Boulder: Westview.

Clark, T. N., & Rempel, M. (Eds.), (1997). *Citizen Politics in Post-Industrial Societies*. Boulder: Westview.

Contarino, M. (1995). The Local Political Economy of Industrial Adjustment: Variations in Trade Union Responses to Industrial Restructuring in the Italian Textile-Clothing Sector. *Comparative Political Studies 28*, 62–86.

Cox, C., & Young, W. (1995). Compulsory Competitive Tendering in Northern Ireland Local Government. *Local Government Studies 21*, 591–606.

Danziger, J. N. (1980). California's Proposition 13 and the Fiscal Limitation Movement in the US. *Political Studies 29*, 599–612.

DeFilippis, J. (1999). Alternatives to the New Urban Politics: Finding Locality and Autonomy in Local Economic Development. *Political Geography 18*, 873–990.
Dicken, P. (1998). *Global Shift: Industrial Change in a Turbulent World.* Third Edition. London: Harper and Row.
Doherty, B. (1998). Opposition to Road Building. *Parliamentary Affairs 51*, 370–383.
Downs, A. (1957). *An Economic Theory of Democracy.* New York: Harper and Row.
Duncan, S. S., & Goodwin, M. (1988). *The Local State and Uneven Development.* Cambridge: Polity.
Dunleavy, P. J. (1986). Explaining the Privatization Train. *Public Administration 64*, 13–34.
Dye, T. R., & Robey, J. S. (1980). Politics Versus Economics: Development of the Literature on Policy Determination. In: T. R. Dye, & V. Gray (Eds.), *The Determinants of Public Policy,* (3–17). Lexington, Massachusetts: DC Heath.
Elazar, D. J. (1970). *Cities of the Prairie.* New York: Basic Books.
Elazar, D. J. (1986). *Cities of the Prairie Revisited.* Lincoln: University of Nebraska Press.
Evans, C. (1985). Privatization of Local Services. *Local Government Studies 11*(6), 97–110.
Friedmann, J. (1995). The World Cities Hypothesis. In: P. L. Knox, & P. J. Taylor (Eds.), *World Cities in a World System,* (317–331). Cambridge: Cambridge University Press.
Fuchs, E. (1992). *Mayors and Money.* Chicago: University of Chicago Press.
Fukuyama, F. (1995). *Trust.* New York: The Free Press.
Garofoli, G. (Ed.). (1992). *Endogenous Development and Southern Europe.* Aldershot: Avebury.
Giddens, A. (1984). *The Constitution of Society: Outline of the Theory of Structuration.* Berkeley: University of California Press.
Griggs, S., Howarth, D., & Jacobs, B. (1998). Second Runway at Manchester. *Parliamentary Affairs 51*, 358–369.
Guterbock, T. M. (1980). *Machine Politics in Transition: Party and Community in Chicago.* Chicago: University of Chicago Press.
Hall, T., & Hubbard, P. (Eds.). (1998). *The Entrepreneurial City.* Chichester: Wiley.
Harding, A. (1991). The Rise of Urban Growth Coalitions – U.K. Style?. *Environment and Policy C: Government and Policy 9*, 295–317.
Harding, A., Dawson, J., Evans, R., & Parkinson, M. (Eds.). (1994). *European Cities Towards 2000: Profiles, Policies, and Prospects.* New York: St Martin's Press.
Harvey, D. W. (1989). From Managerialism to Entrepreneurship: The Transformation of Urban Governance in Late Capitalism. *Geografiska Annaler 71B*, 3–17.
Heller, P. (1995). From Class Struggle to Class Compromise: Redistribution and Growth in a South Indian State. *Journal of Development Studies 31*, 645–672.
Henney, A. (1984). *Inside Local Government: A Case for Radical Reform.* London: Sinclair Browne.
Hirst, P., & Thompson, G. (1996). *Globalization in Question.* Cambridge: Polity.
Hoggart, K. (1987). Does Politics Matter?: Redistributive Policies in English Cities 1949–74. *British Journal of Political Science 17*, 359–371.
Inglehart, R. (1990). *Culture Shift.* Princeton, New Jersey: Princeton University Press.
Kaase, M., & Newton, K. (1995). *Beliefs in Government.* Oxford: Oxford University Press.
Klingemann, H-D., & Fuchs, D. (Eds.). (1995). *Citizens and the State.* Oxford: Oxford University Press.
Law, C. M. (1993). *Urban Tourism.* London: Mansell.
Listhaug, O., & Wiberg, M. (1995). Confidence in Political and Private Institutions. In: H-D. Klingemann, & D. Fuchs (Eds.), *Citizens and the State,* (298–322). Oxford: Oxford University Press.
Lorrain, D., & Stoker, G. (Eds.). (1997). *The Privatization of Urban Services in Europe.* London: Pinter.

McAdam, D., McCarthy, J. D., & Zald, M. N. (Eds.). (1996). *Comparative Perspectives on Social Movements*. Cambridge: Cambridge University Press.
Mann, M. (1986). *The Sources of Social Power – Volume One*. Cambridge: Cambridge University Press.
Miranda, R. (1994). Containing Cleavages: Parties and Other Hierarchies. In: T. N. Clark (Ed.), *Urban Innovation*, (79–103). Thousand Oaks, California: Sage.
Mullins, D., Niner, P., & Riseborough, M. (1995). *Evaluating Large Scale Voluntary Transfers of Local Authority Housing*. London: HMSO.
Nolan, J. L. Jr. (1998). *The Therapeutic State: Justifying Government at Century's End*. New York: New York University Press.
Patterson, A., & Pinch, P. L. (1995). Hollowing Out the Local State: Compulsory Competitive Tendering and the Restructuring of British Public Sector Services. *Environment and Planning A27*, 1437–1461.
Perry, D. C. (Ed.). (1995). *Building the Public City: The Politics, Governance and Finance of Public Infrastructure*. Thousand Oaks, California: Sage Urban Affairs Annual Review 43.
Peterson, P. E. (1981). *City Limits*. Chicago: University of Chicago Press.
Pierre, J. (Ed.). (1998). *Partnerships in Urban Governance: European and American Experience*. Basingstoke: Macmillan.
Pinch S. P. (1989). The Restructuring Thesis and the Study of Public Services. *Environment and Planning A21*, 905–926.
Piore, M., & Sabel, C. (1984). *The Second Industrial Divide*. New York: Basic.
Putnam, R. (1994). *Making Democracy Work*. Princeton: Princeton University Press.
Putnam, R. (1995). Bowling Alone. *Journal of Democracy* 6(1), 6578.
Rider, C., & Zajicek, E. K. (1995). Mass Privatization in Poland: Processes, Problems and Prospects. *International Journal of Politics, Culture and Society* 9, 133–148.
Royal Commission on Local Government in England (1969). *Report*. London: HMSO.
Saiz, M., & Geser, H. (Eds.). (1999). *Local Parties in Political and Organizational Perspective*. Boulder: Westview Press.
Saltzstein, A. L. (1986). Did Proposition 13 Change Spending Preferences?. In: T. N. Clark (Ed.), *Research in Urban Policy: Volume 2A – Fiscal Austerity and Urban Management*, (145–157). Greenwich, Connecticut: JAI Press
Scott, A. J. (1988). *New Industrial Spaces: Flexible Production Organization and Regional Development in North America and Western Europe*. London: Pion.
Sellgren, J. M. A. (1991). The Changing Nature of Economic Development Activities: A Longitudinal Analysis of Local Authorities in Great Britain, 1981-87. *Environment and Planning C: Government and Policy* 9, 341–362.
Shefter, M. (1985). *Political Crisis/Fiscal Crisis: The Collapse and Revival of New York City*. New York: Basic Books.
Smith, M. P. (1995). The Disappearance of World Cities and the Globalization of Local Politics. In: P. L. Knox, & P. J. Taylor. (Eds.), *World Cities in a World System*, (249–266). Cambridge: Cambridge University Press.
Susskind, L. E., & Serio, J. F. (1983). *Proposition 2 ½: Its Impact on Massachusetts*. Cambridge, Massachusetts: Oelgeschlager Gunn and Hain.
Swianiewicz, P., Blaass, G., Illner, M., & Péteri, G. (1996). Policies: Privatizing, Defending the Local Welfare State, or Wishful Thinking?. In: H. Baldersheim, M. Illner, A. Offerdal, L. Rose & P. Swianiewicz (Eds.), *Local Democracy and Processes of Transformation in East-Central Europe*, (160–196). Boulder: Westview.
Swyngedouw, E. A. (1995). The Contradictions of Urban Water Provision: A Study of Guayaquil, Ecuador. *Third World Planning Review* 17, 387–405.

Tabb, W. K. (1984). The New York City Fiscal Crisis. In: W. K. Tabb, & L. Sawers (Eds.), *Marxism and the Metropolis,* (323–345). Second Edition. New York: Oxford University Press.

Thompson, M., Ellis, R., & Wildavsky, A. (1990). *Cultural Theory.* Boulder: Westview.

Urry, J. (1986). Class, Space and Disorganized Capitalism. In: K. Hoggart, & E. Kofman. (Eds.), *Politics, Geography and Social Stratification,.* 16–32. London: Croom Helm.

Waters, M. (1995). *Globalization.* London: Routledge.

2. THE LEGITIMACY OF LOCAL GOVERNMENT – WHAT MAKES A DIFFERENCE? EVIDENCE FROM NORWAY[1]

Lawrence E. Rose and Per Arnt Pettersen

> On both methodological and substantive grounds . . . the subject of legitimacy must count as one of the central issues of social science.
>
> (Beetham, 1991, p.7)

Observers of local government in Norway looked forward to election night on the 13th of September 1999 with some trepidation. The reason was not widespread superstition relating to the number 13. Rather, grounds for apprehension lay in the election outcome four years earlier. In 1995 little more than 62% of eligible voters cast a ballot in municipal elections. Even less cast ballots in county council elections (59%).[2] While this level of voting may be respectable by international standards, in Norway it gave rise to sharp concern. It was, after all, the lowest level recorded in the entire postwar period. One has to go back to 1922 to find a lower turnout in municipal elections for the nation as a whole.[3] Every bit as noteworthy was the fact that this result strengthened a trend that has prevailed since the historical high for turnout at local elections (81%) in 1963 (see Figure 2.1).[4] In a little over 30 years, the percentage of those who chose to pass up the opportunity to vote in local elections has roughly doubled (from 19% to 38%).

The electoral outcome in 1999 did little to allay these apprehensions. On the contrary, voter turnout continued to decline, falling to 60% for municipal

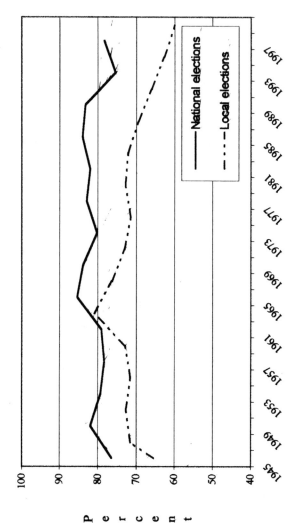

Figure 2.1: Voter Turnout in National and Local Elections in Norway, 1945-1999.

Source: Statistik sentralbyrå (1996, p.22; 1998, p.28).

elections and just above 56% for county council elections. Efforts to lower the electoral threshold by permitting postal balloting prior to election day, and to stimulate electoral participation, including experiments with the direct election of mayors in 20 of Norway's 435 municipalities, appeared to be largely for naught – or to have had marginal effects at best. The trend of declining electoral involvement continues unabated.

This trend has been disquieting for many, especially for politicians who experience a pronounced need, or at least find it convenient, to anchor their actions in a mandate from 'the people's voice'. In a post-election commentary in 1995, for example, Bengt Martin Olsen, the retiring mayor of Bergen, Norway's second largest municipality, proclaimed:

> In my opinion, democracy can't survive with an electoral turnout of 50 to 60%. Under such conditions there are too few people who decide matters. It is a defeat [for local democracy] if we can't get the voters to be engaged in local affairs.
> (Cited in *Aftenposten*, 14 September 1995, translated by the authors.)

Decline in turnout not only weakens the moral strength of politicians' mandates, but in a more general sense calls into question the legitimacy of local government. This is because elections are seen by many to constitute a keystone of democracy. A decline in their use raises doubts about the foundation upon which democracy is built. Hence, even for a distanced observer, it is necessary to ask how long-term decline in electoral turnout is to be interpreted and what implications, if any, it has for the legitimacy of local government?[5]

Several responses, each of which is logically feasible, present themselves. One possibility, which is in keeping with the 'dire straights interpretation' offered by many political figures, is that Norwegian local government is suffering from a long-term decline in legitimacy, and declining electoral turnout is but one manifestation of this. Such an interpretation may be supported by the problems encountered by many local party organizations in finding local election candidates and in the blossoming number of local candidate lists that operate independently of the regular national parties. These local lists frequently involve an element of 'protest', reflecting the disenchantment of segments of the population with the functioning of local government (cf. Berglund & Bjørklund, 1994, pp. 15–18; Arstein, 1996; Bjørklund, 1999, p. 124). To be sure, protests may focus as much on specific local issues or the constellation of local political forces as on the underlying system of government. However, in the absence of more explicit empirical evidence, the possibility of genuine frustration with and diminished legitimacy of local government cannot be precluded.

An alternative and more benign interpretation of electoral events presents almost a polar opposite. This is that falling turnout signals a sense of contentment, an assessment that 'all's well' on the local home front; there are no burning issues that serve to mobilize the electorate (cf. Wilson, 1936; Lipset, 1963, p. 217). From this perspective low turnout suggests that local government has for the most part lived up to expectations and that there is no impending crisis of legitimacy. Rather a wellspring of general confidence exists amongst the population.[6]

Other possibilities may be entertained as interpretations of recent electoral results. Given that turnout in national elections showed similar decline before rebounding slightly in the most recent parliamentary election (see Figure 2.1), it is possible that what is happening may be no more than a 'natural' swing of events in the life of a political system.[7] That said, our intent in this chapter is not to draw forth, investigate and render judgment on all possible interpretations of declining electoral participation. Our intent is to investigate citizen perceptions that arguably relate to the legitimacy of local government in Norway. The investigation is based upon panel data drawn from two national surveys carried out in 1993 and 1996, respectively (cf. Rose & Skare, 1996a, 1996b).

In what follows we first devote a few words to the concept of legitimacy and the manner in which we operationalize this concept in our analyses. We then review previous empirical studies of political legitimacy to identify factors that may explain variations in perceptions of local government institutions. Following this we present our results and conclude the chapter with a brief discussion of our findings.

THE CONCEPT OF LEGITIMACY

As the opening citation from David Beetham (1991) suggests, the subject and concept of legitimacy occupies a prominent position in the pantheon of moral, legal and political issues. The attendant literature is voluminous. Exegeses and definitions of the term abound, virtually all of which recognize and serve to highlight the complex, multi-faceted character of the concept. The work of Max Weber (1964) constitutes a central benchmark in this regard.[8] The essence of Weber's definition is that legitimacy reflects a psychological disposition; it is a belief or perception that the manner in which power and authority are exercised are in accordance with certain norms – whether these norms are traditionally, charismatically or legally-rationally defined. In most contemporary democracies, legal-rational norms have largely replaced traditional and/or charismatic foundations for perceptions of legitimacy, although traces of both, especially of charismatic authority, certainly remain. Norway is no exception in this respect. But it is appropriate to emphasize that in Norway, much as for all the Nordic countries, legal and bureaucratic principles have deep and steadfast roots, being anchored in a culture characterized by over-arching concerns with equality, democracy, and the rule of the law (cf. Eckstein, 1966; Graubard, 1986).

For the present analysis, therefore, our point of departure is how the populace perceives a number of actors or objects that occupy a salient position within democratic local self-government.[9] In particular, we direct attention to perceptions of four specific objects: (1) local politicians; (2) local parties; (3) the municipal council and its leader (the mayor); and, (4) the local administration. These four do not exhaust potentially relevant sources for perceptions of local government legitimacy, but all are (as the following thumbnail remarks should indicate) critical to the operation of democratic local government in Norway, much as in many other countries. *Politicians* are those who most directly participate in debates over local agendas, issues and priorities. *Parties* are the organizations that aim to aggregate the public interest into policy solutions that may be presented to the public through electoral platforms and campaigns. *The municipal council* (a body elected by proportional representation and vested with penultimate authority for local decision-making) is supposed to respond to public preferences and inspire confidence in local self-government. Finally, *the local administration* is the agency responsible for preparing the groundwork for political decisions, as well as implementing policy decisions taken, in a predictable, even-handed and efficient fashion.

Viewed in terms of David Easton's (1965) analytical framework, all four of these objects are part of the input and conversion processes of local government. Although it is reasonable to suggest, as Easton's model implies, that policy outputs may similarly be 'objects' whose perception by the public may strengthen or weaken local political legitimacy, we prefer to treat the perception of outputs (i.e. an individual citizen's satisfaction or dissatisfaction with general policy outputs) as an independent variable. In so doing, the idea is to investigate whether policy outputs and the perceptions they generate have feedback into the perception of objects in the input and conversion process, as Easton's model suggests.

In the analysis undertaken here, we are not in a position to investigate the legitimacy of local government directly and rigorously. That is, we are not able to establish the precise *degree or level* of legitimacy ascribed to local government *per se*, or the specific grounds people have for ascribing legitimacy to local government. Such an analysis would require probing individual expectations, norms or value orientations regarding a wide variety of objects of local government, as well as exploring how far these are perceived to be satisfied. It would then require considering what implications any possible discrepancy might have for individual perceptions about the legitimacy of a particular set of political arrangements.

The strategy we pursue is of a more indirect character. We employ responses to a variety of questions included in two national surveys, in which questions were designed to tap perceptions of *different properties* relating to the four principal objects of local government already identified, as well as the general manner in which local government functions. One such property, which has been central to many investigations, concerns the general sense of *trust or confidence* regarding specific actors or institutions. In addition the surveys contained questions relating to the *competence* of politicians, the *indispensability* of political parties, and the *responsiveness* of the municipal council.[10] Again these properties are not exhaustive, but they capture components that are relevant to a multi-faceted concept of political legitimacy.[11]

It is appropriate to emphasize that considering the legitimacy of local government in this fashion focuses attention on that which, in Easton's terms, constitutes aspects of 'diffuse support' for *specific authorities*. We are to a lesser degree focusing on the local regime, and we do not consider perceptions of the legitimacy of the *political community* more generally. As Easton and others have noted, specific authorities may well lack support or legitimacy, but the regime and/or political community of which they are part may remain intact, as it is held in high regard by the same individuals who hold negative views of specific political-administrative office holders.[12] In short, the principal thrust in our analysis is to consider what factors may contribute to positive perceptions of specific local authorities and, to some extent, the local regime, but not the political community *per se*. The underlying presumption is that such perceptions have a strong bearing on the legitimacy of local government.

EXPLAINING POSITIVE PERCEPTIONS: THEORETICAL EXPECTATIONS

A review of empirical studies of political legitimacy during the last few decades reveals at least three clear tendencies. First, as their measure of political legitimacy, a significant majority use public confidence or trust in political actors or institutions. In particular politicians and political parties are in the limelight, although a good number of studies also look at confidence in other social, economic and political institutions as a benchmark.[13] By contrast, measures of more fundamental aspects of political legitimacy – especially those relating to what Easton terms the political regime and the political community respectively – are less commonly investigated.[14]

Second, many empirical studies find evidence of a decline in confidence and trust in politicians and political parties amongst the mass public. Based on such evidence, some of the scholarly literature, and even more so the non-scholarly literature, suggests that democratic societies are facing secular decline in legitimacy that may pose serious problems in the long-term (cf. Crozier et al., 1975; Lipset & Schneider, 1983). Closer inspection of the data, however, reveals that this trend is by no means universally valid. In fact expressions of confidence or trust may vary in keeping with politically significant events. Evidence from Norway covering the period from 1973 to 1995, for example, showed increased expressions of public confidence in politicians following the referenda on Norwegian membership of the European Community, although such assessments dropped markedly from 1985 to 1989.[15] Comparative evidence from the four European countries reviewed by Listhaug (1995) tends to support the view that shifts in public confidence are a reflection of short-term 'political realities', although there are longer term declines in some cases.[16]

A third tendency, which is perhaps most relevant for the present investigation, is that prior studies of political legitimacy predominantly focus on the national level. To be sure, studies of local or sub-national political legitimacy are to be found (e.g. Agger et al., 1961; Nielsen, 1981; Baldassare, 1985; Andersen et al., 1992; Kornberg, 1992; Nordtun, 1992; Sørensen, 1996), but their number is limited in comparison with those at the national level. In one respect this is understandable. The nation state has long constituted the principal frame of reference for much work in political science particularly among those concerned with the building and maintenance of national political systems. There is a certain logic to this focus, inasmuch as the nation has, at least until recently, provided an important context within which other political bodies and activities have their *raison d'être*. All the same, the lack of studies concerning the legitimacy of local government is surprising. This is particularly so given strong growth in the role played by local government in many democratic countries (cf. Sharpe, 1988). Local governments, especially in the Nordic countries, have come to occupy a highly salient position in public service provision, as reflected in the percentage of public consumption and employment located at the sub-national level (cf. Albæk et al., 1996).[17] For practical as well as theoretical reasons, therefore, there seems to be a clear justification for more work on perceptions of local government legitimacy.

An obvious question in this regard is whether the theoretical propositions advanced with respect to political legitimacy at the national level are equally appropriate or valid at the local level. Our initial supposition is that whereas local and national political legitimacy share features in common, it is reasonable to presume that they are formed and exist independently of one another. Confidence or trust in local political actors and institutions may well increase at the same time as confidence in national political actors and institutions decline, and vice versa. The manner in which various factors contribute to either the development or undercutting of political legitimacy at each level may also be different. In short, much work, both theoretical and empirical, remains to be done.

Rather than starting from scratch and 'inventing the wheel anew', we begin with explanatory propositions generated in previous studies of national political legitimacy, extending and supplementing these as appropriate. Propositions in the extant literature may be roughly sorted into two major categories. On the one hand, there are macro-level secular trend interpretations suggesting a long-term decline in political legitimacy. On the other hand, micro-level and contextually based interpretations to a greater degree suggest that fluctuations in political legitimacy occur, particularly with respect to short-term support for specific political actors and institutions. Although we concentrate our attention on factors in the second category, both types of explanations are reviewed briefly.[18]

Macro-level 'Secular Trend' Arguments

At least three distinct arguments suggest the significance of macro-level secular trends in relation to perceptions of political legitimacy. The first may be termed the *overload hypothesis*. This argument, which emerged and was quite prevalent in the 1970s, has its roots in the major expansion of public sector activity in many countries in the postwar period, especially in connection with welfare services.[19] According to this argument, expansion generated expectations which, when governments were subsequently forced to make adjustments in the face of new realities, produced negative public assessments and a decline in political legitimacy.

This thesis gains additional significance when coupled with arguments suggesting a general shift in citizen orientations. Such an argument is perhaps most closely associated with the work of Jürgen Habermas (cf. 1994), but either implicitly or explicitly it is intertwined in the work of many other writers.[20] At issue is whether the welfare state, and in particular the proliferation of welfare programs based on principles of universal entitlement, may have contributed to a shift in the way individuals define their relationship to the public sphere and hence evaluate government. The underlying hypothesis is that a shift has occurred, or at least is likely to occur. The suggestion is that, given a relatively encompassing set of social as well as civil and political rights, individuals are inclined to turn away from their role as *citizen-voters* to focus more narrowly on their role as *citizen-consumers*, seeking to secure the public goods and services to which they and their closest family members are entitled. Such a shift may be enhanced in situations where governments, through various reform measures, many under the guise of a New Public Management philosophy, encourage citizens to think of themselves as customers.

Without better longitudinal data, it is difficult to assess whether such a shift has taken place. What evidence exists for Norway does suggest that individuals tend to favor values of effectiveness in municipal service delivery more strongly than values of local democracy and autonomy. Moreover, they emphasize their role as consumers of services more than their role as taxpayers or politically engaged citizens (Pettersen & Rose, 1997; Rose, 1999).[21] To the extent such attitudes are prevalent, whether due to a recent shift or not, it is reasonable to expect cutbacks in public programs to meet with irritation and even outrage, particularly among those with a pronounced citizen-consumer orientation, and that perceptions of legitimacy would suffer accordingly.

A second secular trend argument is based on the *'new politics'* hypothesis. This hypothesis is most clearly associated with the theoretical thinking advanced by Ronald Inglehart and Terry Clark (cf. Inglehart, 1977, 1990, 1997a, b; Rempel & Clark, 1997; Clark & Inglehart, 1998). According to this argument, voters in recent years place greater emphasis on a new issues or priorities, many of which are insufficiently reflected in the policies of contemporary political systems. To the extent this discrepancy exists and may grow over time, especially as the proportion of voters with new value orientations grow, it is reasonable to expect increased disenchantment and declining expressions of support for government.

This argument is particularly relevant when coupled with the rise in education levels among the mass public. Education increases the potential for individual political competence and hence for independence of citizens from actors or institutions that traditionally served as intermediaries between the individual and the political system. It is suggested, for example, that with increased education, relative dependence upon and the importance of political parties as a source of cues for thinking about political life has declined.[22] On the surface, of course, education is neutral. It need not automatically lead either to positive or to negative change in perceptions of political legitimacy. But one may nonetheless suggest that education, by its very nature, may imply a bias. It presumably generates a 'healthy' skepticism to fixed or pre-packaged answers, and prompts a more critical stance toward mediating institutions like political parties, the mass media or other opinion leaders. If so, this could have negative consequences for a sense of trust or confidence in political actors and institutions – politicians and political parties in particular.

The third secular trend argument relates to practices in the mass media. This may be termed the *'negative journalism'* hypothesis. According to this hypothesis, mass media journalism has shown a marked tendency to focus on negative news. This 'negativity bias' provides an excess of derogatory information about political actors and events. To the extent this information is consumed and integrated into assessments of political objects, this bias colors perceptions of political legitimacy among the mass public. This hypothesis runs parallel to the suggestion of Gamson (1968) that much of the decline in political trust, along with the increased political cynicism observed at the end of the 1960s, largely resulted from an accumulation of dissatisfaction which in many instances was directed toward 'innocent' political figures.[23] Such an interpretation is in keeping with a popular belief among politicians that falling trust and increased cynicism relating to their position is largely media-generated; that they are victims of a 'bad rap'. Numerous examples of this belief are to be found (cf. Lipset & Schneider, 1983, p. 406), and much attention has been devoted to this issue in the academic literature (cf. Robinson, 1976; Miller et al., 1979; Lau, 1982, 1985; Iyengar & Kinder, 1987). The validity of the general argument, however, is subject to question. As Listhaug (1995, p. 264) noted: ". . . blaming the media for a decline in political trust rests on a set of assumptions which are not easily verified."

Micro-level and Contextual Arguments

A second set of theoretical arguments place more emphasis on the manner in which citizens, based on their individual circumstances, experience 'political reality'. The potential relevance of such views is obvious. People experience the same macro-level phenomena differently and react accordingly, thereby giving rise to individual variations in perceptions of political legitimacy. Thus, it is entirely proper that shorter-term, micro-level factors should be considered in seeking explanations of variation of political legitimacy.

Significantly, previous micro-level studies of political legitimacy have turned up little evidence that mainstream socio-demographic factors, like *age, gender and civil status*, which figure prominently in investigations of political attitudes and behavior, are noteworthy discriminators. What findings exist are rather weak and mixed. Agger and associates (1961, p. 488) and Litt (1963, p. 316), for example, found that political confidence tended to decline with age, whereas Lipset & Schneider (1983, p. 122) found a weak tendency in the opposite direction. This latter finding is in keeping with the idea that an individual's stake in and identification with society (and hence the likelihood of perceiving institutions more favorably?) increase as a function of time. One might likewise imagine that variations in age, gender and civil status reflect differences in dependency on public services (it is frequently suggested, for instance, that women tend to be more dependent on public services than men, especially with an increasing prevalence of single parent families), which should impact on perceptions of public authorities, but there is little empirical support for this.

By contrast, socio-demographic factors like *education* and *income* have more frequently been found to be of significance, although again the findings aremixed. Early studies by Agger and associates (1961, p. 484) and McDill & Ridley (1962, p. 211) found those with higher education expressed more confidence in political actors and institutions. In more recent work, however, education has often been found to be negatively related to expressions of political legitimacy. The argument advanced to explain these negative findings is, as suggested previously, that education increases political competence and a sense of skepticism, which gives rise to more critical assessments of political figures and institutions (cf. Listhaug & Wiberg. 1995, p. 314).[24] Against this argument it may be suggested that political competence acquired through or facilitated by education does not necessarily lead to a cynical orientation *per se*, but to a greater degree of *political realism*. Realism may, depending upon the circumstances that pertain, produce either a negative or a positive assessment of the performance of political actors and institutions. As an example, realism that gives an appreciation of the need to make tough decisions in situations of multiple constraints by no means automatically leads to more negative evaluations. The question is under what circumstances is realism most likely to lead to positive assessments and perceptions of political legitimacy rather than the opposite.[25]

Another socio-demographic characteristic that is less frequently investigated is the individual's *occupational position* – in particular whether a person works in the public or private sector.[26] The most reasonable expectation here is based on the presumption that those employed in the public sector have a vested interest. If so, they should characteristically be more inclined to render positive judgments regarding other public sector actors and the institutions within which they operate. One may object, of course, that precisely because of their regular contact with politicians and political institutions, public sector employees should possess greater insight and realism than those 'outside the system'. Potentially this realism may cut both ways. All the same, it seems most likely that a 'rational' self-interest will prevail and lead to predominantly positive assessments.

Other individually-based factors that quite commonly have been found to be relevant to perceptions of political legitimacy relate to subjective orientations. Prominent in this regard are *personal self-confidence*, a *sense of political competence*, and a *sense of trust in others*. The argument here is that perceptions of political legitimacy are in large measure an extension of general psychological dispositions toward the self. As Robert Lane (1959, p. 164) has stated:

> If one cannot trust other people generally, one can certainly not trust those under the temptations of and with the powers which come with public office. Trust in elected officials is seen to be only a specific instance of trust in mankind.

Analyses by Lane and many others support this general proposition (cf. Agger et al., 1961, p. 490; McDill & Ridley, 1962, p. 206; Almond & Verba, 1963, p. 285). In a similar fashion, a sense of personal inadequacy, both in general and with respect to political affairs, may have an 'infectious impact' on perceptions of political legitimacy. By contrast, those who feel more politically knowledgeable and efficacious, who are interested and actively engaged in politics (e.g. as party members) may view the political system more favorably.

Prominent among arguments relating to micro-level factors with an explicit political content are those concerning *ideological considerations*. Here several strands are evident, most of which tie into the concept of 'political distance', a concept first introduced by Miller (1974). The principal thrust of the argument is that political support and perceptions of political legitimacy will be a function of the distance between an individual's ideological preferences and the stands articulated and adopted by political actors and institutions. Hence, voters who support the party or parties in office are likely to be more positively inclined than those who support parties that do not hold office. Similarly, regardless of partisan preferences, those holding views in keeping with prevailing public policy should express more positive evaluations than those who oppose existing policies. Stripped of its partisan ideological content and preferences for specific policy options, this argument may be extended to suggest that those who are more satisfied with policy outputs and the performance of public institutions in general will be most likely to express support for and confidence in politicians and public authorities. In other words, we should expect those who, for whatever reason, give high marks to municipal activities and public service delivery to be more favorably disposed to political actors and/or institutions than those who are dissatisfied.

What, in addition to partisan orientations, may contribute to satisfaction with municipal activities is a topic for extensive discussion in its own right. Conditions relating to individual needs and objective service provision would seem to be obvious, but empirical findings provide only mixed support for such a notion (cf. Schuman & Gurenberg, 1972; Birgersson, 1977; Gaziel, 1982). Equally important may be personal orientations and priorities with respect to municipal activities. For example, variation in the degree to which individuals place emphasis on the municipality as an arena for democratic citizen involvement, as an agent for the collection and use of fees and taxes, or as an efficient producer of public goods and services, could give rise to differences in satisfaction with municipal activities and hence with the political actors and/or institutions which most prominently represent the municipality. Given variation in the resource base for local service performance, one should, on the basis of the general drift of the argument here, expect local variation in expressions of political support and perceived political legitimacy – variations that are independent of longer-term secular trends.

The Relevance of Municipal Size

The theoretical perspectives reviewed up to this point are largely developed and investigated in the context of national political actors and institutions. It is therefore no surprise that little systematic attention has been devoted to variation in political support and legitimacy across political-administrative units within a political system, or to the relevance of conditions pertaining to these units. Among the variety of conditions which may be of theoretical relevance, we highlight one in particular – namely, municipal size.

There are at least two reasons for suggesting that municipal size may be important. The first is of a social-psychological character. The argument is that smaller units of government tend to be perceived more favorably than larger units; they are perceived as being 'nearer-at-hand', more personable and friendly than larger units, which *ipso facto* are more distant, impersonal and bureaucratic. Put simply, smaller scale government is more readily grasped and fathomed than larger authorities, and this familiarity is likely to advantage the actors and institutions of smaller authorities. This line of reasoning runs counter to the notion that 'familiarity breeds contempt', but there is little reason to believe this notion is valid for perceptions of political legitimacy. On the contrary, there is good reason to believe the opposite. In smaller units of government the likelihood that citizens know politicians, public figures and employees personally is greater. These people may be neighbors or even relatives, and residents in smaller units of government have more direct access to these officeholders, being able to speak with them directly, or at least follow the workings of government more closely (from the gallery, so to speak). Smaller units of government, in short, are more transparent and permeable, so local residents are more likely to feel a greater sense of personal efficacy.

This line of thinking lies at the heart of Dahl & Tufte's (1973, p. 20ff) discussion on citizen effectiveness in *Size and Democracy*. The relationships suggested here are not automatic, nor are they necessarily linear. There may be threshold and ceiling effects with respect to size, such that the beneficial consequences of smallness dissipate rapidly with increasing size. Whether this is the case, and at what point the 'law of diminishing returns' sets in, is a matter of empirical inquiry. But at least in a political system such as Norway, where there are many small units of local government, we expect to find evidence of such a tendency.[27]

Within the Norwegian political system, moreover, the issues relating to the size of local authorities and possible amalgamation reforms have been matters of nearly continuous public debate throughout the postwar period.[28]

Our second reason for suggesting the importance of municipal size is based more on an information flow argument. For national government, the flow of political information tends to be dominated by central sources – national newspapers, national television and radio, and the political messages of national political parties. These sources are commonly available to one and all, and easily become the 'property' of any politically interested citizen, regardless of physical location. By contrast, political information regarding local government is more geographically differentiated. The smaller the unit of government, the more local information is likely to have a specific, uniquely identifiable component that allows alert citizens to relate it to their own place of residence, thereby permitting them to render more distinct judgments with a bearing on perceptions of political legitimacy. This argument is particularly relevant in a country like Norway, which has a highly differentiated local media (cf. Høst, 1991).

To summarize, to the extent that proximity and visibility, as well as the availability of more differentiated information, may be argued to have an impact on citizen assessments of political actors and institutions, we would expect confidence relating to *local* political actors and institutions to be greater among residents of smaller municipalities.

Other theoretical propositions may be entertained, some possibly with greater relevance.[29] Our objective, however, is not to be exhaustive. Rather, we seek to explore the explanatory power of factors that have received attention in studies of perceptions relating to political actors or objects at the national level. The intent is to assess how relevant these factors may be at the local level. With this thought, let us turn to the empirical evidence.

VARIATIONS IN CONFIDENCE – WHAT MAKES A DIFFERENCE?

Our empirical investigation is designed to assess the relevance of micro-level and contextual factors in a systematic fashion. We employ standardized least squares regression analysis, using 15 explanatory variables against eight indices reflecting different properties of the political objects investigated. Presentation and discussion of our findings is structured accordingly. The first results to be presented pertain to confidence or trust in *politicians*. These are compared to perceptions of how politicians perform. Next follows the question of whether *parties* are necessary in local government. These findings are contrasted with perceptions of how parties perform. After this comes confidence in the *municipal council*, together with perceptions of the responsiveness of the municipal council. Lastly, the analysis turns to confidence in the *local administration* and to perceptions of the general performance of local government.

Confidence in Local Politicians:
Greatest Among Those 'in the Know' and the Satisfied

Normally questions about trust in politicians are formulated in a fairly general way: A typical question takes the form 'How much confidence do you have in politicians in ...'?, after which follows the name of some political unit or community and a measurement of sympathy. Such a formulation could disguise a huge difference between asking about politicians in general, and those politicians belonging to specific parties – especially the party of a respondent's own choice. We avoid this possible pitfall by posing two questions, one about confidence in politicians in the party the respondent feels nearest to ideologically, the second about confidence in politicians in all other parties. Responses to the two questions show substantial variation, but are sufficiently correlated ($r = 0.50$) that they are combined into an index of 'confidence in local politicians'.

Results in the first column of Table 2.1 reveal that confidence in local politicians is not highly concentrated within specific population groups. Moderate regression coefficients and a low adjusted R^2 indicate that confidence varies rather weakly with the indicators we have selected, even if some significant exceptions are evident. Most prominent in generating positive perceptions of confidence in local politicians is a personal sense of being well informed about local political affairs and general satisfaction with local services. By comparison, those who score high on the index for personal concern with how the municipality finances local activities and uses tax money tend to hold more negative views. This relationship corresponds with household income: those in the highest income brackets have the least confidence in local politicians. In keeping with the arguments suggested previously, there is also evidence that residents of smaller municipalities are more favorably disposed to local politicians. A bit surprising perhaps is that older individuals express less confidence in local politicians than younger persons. This is a relationship to which we will return.

Performance of Local Politicians: The Satisfied and Those in Small
Municipalities are Most Positive

The performance of local politicians – whether they do well or badly – may be measured in many ways. Here three standard performance measures are used:

Table 2.1 Regression Coefficients for Perceptions Regarding Confidence in Local Politicians and their Performance

Independent Variables	Confidence in Local Politicians	Performance of Local Politicians
Age	–0.10**	–0.05
Gender (women)	0.01	0.01
Civil status (married/cohabitant)	0.02	0.00
Education	0.04	0.06*
Household income	–0.07*	0.00
Occupational sector (public)	–0.01	–0.03
Subjective knowledge of local politics	0.14***	0.08**
Subjective local political competence	–0.02	0.18***
Voted for mayor's party (yes)	0.06	0.05
Scope of municipal activity (acceptable)	0.03	0.07**
Citizen role orientations		
Consumer	0.05	0.10***
(Tax)payer	–0.13***	–0.13***
Voter	0.03	–0.05
General satisfaction with municipal services	0.12***	0.23***
Size of municipality (population logged)	–0.08**	–0.22***
Adj. R^2	0.07	0.22
N =	(1,095)	(1,103)

The content of all variables is described in detail in a methodological appendix.
Significance levels: * = 0.05 ** = 0.01 *** = 0.001

do they waste tax money, do they keep their promises, and do they concentrate on re-election rather than the development of their local community. Answers to these three questions are only moderately intercorrelated (coefficients range from 0.13 to 0.34) but are nonetheless combined into an index relating to the 'performance of local politicians' based on their substantive content.[30]

Compared to general confidence in politicians, perceptions of politicians' performance are more closely linked to the explanatory characteristics we consider. The adjusted R^2 is 0.22 in this case (column 2 in Table 2.1). The two variables most strongly related to confidence – subjective knowledge about local politics and general satisfaction with municipal service delivery – continue to be of importance. Personal preoccupation with municipal finance is likewise again associated with more negative assessments of politicians' performance. But in this instance, those preoccupied with their role as consumers of local government services, together with those who find the scope of municipal activity acceptable, express positive assessments of politicians' performance. In addition, both a greater sense of personal political competence and higher formal education (variables which are positively correlated), are associated with a positive evaluation of the performance of local politicians.

These latter relationships are interesting, since they suggest that those who have the strongest foundation for assessing politicians' performance and doing something about it if they be dissatisfied, are the *most* favorably oriented to the performance of local politicians. To look at the 'back side of this coin', negative evaluations appear more likely to flourish among those with lower formal education and a lesser sense of subjective political competence or self-assessed local political knowledge. Although these negative assessments cannot be empirically denied, one can question the foundation upon which the assessments are based. It may be that the adage 'ignorance breeds contempt' is appropriate here.

Perhaps the most noteworthy finding, however, is related to municipal size. The relationship is again negative, but more strongly so, even after other variables are taken into consideration. Everything else being equal, residents of smaller municipalities are much more likely to evaluate the performance of local politicians favorably. This finding is particularly interesting in light of the fact that virtually every proposal put forth by national Norwegian authorities regarding the structure of local government has argued in favor of increasing the size of municipalities. If our findings are valid, such a reform would not strengthen favorable assessments of local politicians, but could be anticipated to have the opposite consequence instead!

Are Local Political Parties Superfluous? Not for Residents of Large Municipalities or the Satisfied

Are political parties really imperative for democratic local self-government? In order to assess opinions on this issue two questions were posed: first, whether local democracy would function just as well without political parties; and, second, if local elections should eliminate the use of party tickets. These items were strongly correlated ($r = 0.51$) and were combined into an index reflecting the 'indispensability of local political parties'.

Remembering that *politicians* are held in highest esteem in small municipalities, it is at first sight a bit surprising that *parties* seem to be held in the lowest esteem in the same municipalities (see column 1 in Table 2.2). But the interpretation of these apparently contradictory findings is actually straightforward. In small municipalities it seems politicians are better known. People know who they are and their basic points of view. They do not need party labels or organizations to determine which politicians to support. There is also good evidence that the individuals preferred by many voters may appear on different party lists. This is documented by a frequently used ballot option

Table 2.2 Regression Coefficients for Perceptions Regarding the Indispensability and Performance of Local Political Parties

Independent Variables	Indispensability of Local Political Parties	Performance of Local Political Parties
Age	–0.07*	–0.16***
Gender (women)	–0.01	0.01
Civil status (married/cohabitant)	–0.02	–0.01*
Education	0.06	0.13***
Household income	–0.05	–0.01
Occupational sector (public)	0.04	0.05
Subjective knowledge of local politics	0.01	0.11**
Subjective local political competence	0.12***	0.22***
Voted for mayor's party (yes)	0.08*	0.07*
Scope of municipal activity (acceptable)	–0.01	0.00
Citizen role orientations		
Consumer	0.06	0.06
(Tax)payer	–0.09**	–0.14***
Voter	–0.01	–0.04
General satisfaction with municipal services	0.12***	0.15***
Size of municipality (population logged)	0.15***	–0.07*
Adj. R^2	0.07	0.24
N =	(1,021)	(928)

The content of all variables is described in detail in a methodological appendix.
Significance levels: * = 0.05   ** = 0.01   *** = 0.001

that allows voters to modify party lists and add the names of individuals who may even appear on other lists. Under such circumstances parties may be considered an obstruction to selecting individuals for positions on the municipal council. In cities and other larger municipalities the situation is rather different. Here few are acquainted with politicians as individuals, and the personal points of view of politicians are hardly known. Under these conditions local political parties function as a useful source of information regarding standpoints. This renders parties more indispensable as political objects.

The juxtaposition of these findings and those in Table 2.1 underline a critical, if obvious, lesson for the study of political legitimacy. This is that the relevance of various conditions as sources of support or legitimacy may vary according to the objects and properties of objects investigated. In the absence of more detailed information such as that provided in Table 2.1, for example, the findings in Table 2.2 could easily lead to the (apparently false) conclusion that if *political parties* are not so popular in smaller municipalities, then the same should also hold for *politicians*. This is not true. In fact, the opposite appears to be the case.

Inspecting the remaining variables, it is those most satisfied with local government services who are most inclined to perceive parties as important for local democracy. These perceptions are stimulated by a sense of local political competence, together with having voted for the party of the mayor. On the other hand, the older generation is not so fond of parties. This also goes for those who are more concerned with how the municipality collects and uses its financial resources.

Performance of Local Parties:
Highest Ratings Amongst the Young and Self-Competent

The next question is who holds positive evaluations regarding the performance of local political parties. More precisely, who believes that parties are concerned with the most important local problems and respond to ordinary people's opinions rather than only being interested in their votes. Responses to these two questions are moderately correlated ($r = 0.27$) and are used to establish an index relating to the 'performance of local political parties'. Results pertaining to this index are found in the second column of Table 2.2.

Here we find another interesting twist. Even if people in smaller municipalities are more inclined to see parties as superfluous, it is in these municipalities that there is a tendency, albeit slight, to perceive parties as competent and responsive. In larger municipalities, where parties are perceived to be more indispensable, their performance is less highly regarded! To the extent there is contempt for politicians and political parties (a notion that is popular in the mass media), it appears that this is most pronounced in larger municipalities.

Positive assessments of local political party performance are also found among those satisfied with municipal service delivery. For both local politicians and local political parties, neither of which are directly responsible for service production and delivery, satisfaction appears to have a beneficial 'spill-over' effect that contributes to more favorable perceptions. Despite this, those who are personally concerned with municipal finance and spending once again remain more critical to the performance of local political parties.

The variable having the strongest impact on favorable perceptions of local party performance, however, is a subjective sense of local political competence. In addition both formal education and a subjective sense of local political knowledge have an important independent effect, even when the impact of other variables is taken into account. In short, those who are competent, both objectively and subjectively speaking, are clearly favorably disposed.

We also see that young people are inclined to perceive the performance of local political parties in a more positive light than their elders. One might be tempted to interpret this as reflecting a certain naiveté bred by 'the ignorance of youth'. But given evidence regarding the impact of formal education, which is greater among the younger generation, as well as subjective local political competence, this cannot be the case. Rather the findings suggest this relationship may be an indication of a generational phenomenon.

Confidence in the Municipal Council:
Greatest Among the Satisfied and Supporters of the Mayor's Party

With this thought, the spotlight shifts to the local decision-making body – the municipal council. The first property of interest here relates to trust or confidence in the institution in general. For this purpose respondents were asked to indicate the amount of confidence they have in the municipal council and its leader (the mayor). Responses to these questions are strongly correlated ($r = 0.59$) and are combined into an index of 'confidence in the municipal council'.

As we see from the first column of Table 2.3, only four variables have a statistically significant impact on relationships of trust, but two are salient. One

Table 2.3 Regression Coefficients for Perceptions Regarding Confidence in and Responsiveness of the Municipal Council

Independent Variables	Confidence in the Municipal Council	Responsiveness of the Municipal Council
Age	–0.00	–0.13***
Gender (women)	0.01	0.04
Civil status (married/cohabitant)	0.02	0.04
Education	–0.03	0.09**
Household income	–0.02	0.03
Occupational sector (public)	–0.00	0.02
Subjective knowledge of local politics	0.08**	0.08**
Subjective local political competence	0.00	0.22***
Voted for mayor's party (yes)	0.22***	0.09***
Scope of municipal activity (acceptable)	0.03	0.09***
Citizen role orientations		
Consumer	0.05	0.04
(Tax)payer	-0.06	–0.13***
Voter	–0.05	–0.09**
General satisfaction with municipal services	0.26***	0.18***
Size of municipality (population logged)	–0.09**	–0.17***
Adj. R^2	0.14	0.26
N =	(1,103)	(1,113)

The content of all variables is described in detail in a methodological appendix.
Significance levels: * = 0.05   ** = 0.01   *** = 0.001

is not very surprising, for those who voted for the mayor's party express more confidence than those who voted for other parties. But notice the decisive impact of those who are satisfied with municipal service provision. By now it seems rather obvious that satisfaction with municipal services is one of the most prominent 'sources' of local government legitimacy. It appears to have an overarching significance above and beyond perceptions regarding specific political objects.

Equally important, however, is the fact that trust or confidence in the municipal council is related to the size of the municipality and a sense of being well informed about local political affairs. Residents in smaller municipalities perceive municipal councils more favorably than those in larger municipalities. Those 'in the know' likewise have more positive assessments than individuals with a lower sense of local political knowledge. But in this case formal education and subjective local political competence do not have a significant impact, nor is there a noteworthy consequence of citizen role orientations. With the exceptions noted, confidence in the municipal council is relatively uniformly spread throughout the population.

Responsiveness of Municipal Councils: Perceived Most Positively by Those with Subjective Political Competence

An essential part of local democracy concerns the ability of elected representatives to comprehend what ordinary people perceive to be important and to respond to the beliefs and attitudes of ordinary people in a measured fashion. The relationship between electors and elected is not only of relevance at election time, but is an on-going affair, in which citizens may evaluate the distance between what they perceive as key issues and the agenda that dominates local politics. Democratic local self-government, in other words, presupposes an element of responsiveness from political institutions regarding common people's views and interests.

Three questions are employed to tap perceptions of this property. These concern whether representatives care for and consider the opinions of ordinary people when decisions are made, if the local council reflects and represents the opinions of most people, and if representatives of the municipal council are able to keep in touch with their voters. Responses to all three questions correlate above the 0.30 level, and are used to construct an index relating to the 'responsiveness of the municipal council'.

The pattern of perceptions evident with respect to this index (column 2 in Table 2.3) is the most intriguing found so far. First, every individual property reflecting some kind of competence – formal education, a self-assessed knowledge of local political affairs and a subjective sense of being competent with respect to local politics – is positively related to a belief that the municipal council is a responsive institution. This may reflect the fact that these groups are those who, by virtue of political participation, influence local politics. If so, this may reflect the fact that those who try to influence local politics are successful, at least some of the time.

Second, those who voted for the party of the mayor perceive local government as responsive. Again this is not surprising, since these individuals probably share many viewpoints in common with the majority of the municipal council. The same argument may be used for those who find the scope of local government activity to be acceptable – our other measure of 'political distance'. Aside from this, there is a rather strong positive impact for general satisfaction with municipal services.

At the same time, a number of variables exhibit a negative relationship with perceptions of municipal council responsiveness. Age and municipal size are two such variables. Older people and residents of larger municipalities are again most critical, while younger people and those in smaller municipalities tend to hold more favorable views. One interpretation of this is that it is the elderly in large municipalities who most acutely experience – subjectively if not objectively – the gap that may emerge between elector and elected in the course of daily politics. Such perceptions may be linked to the fact that local governments in Norway have primary responsibility for care for the elderly, and that questions regarding the ability of some larger municipalities to fulfill this responsibility adequately have frequently been raised in the media in recent years.

That those who emphasize their citizen roles as (tax)payers and voters also tend to hold more negative views of municipal council responsiveness when the impact of other factors is taken into account is also noteworthy. It may be that these people have more pronounced, perhaps even exaggerated and unrealistic, expectations about how responsive the municipal council should be. If so, it would be quite natural for them to experience frustration and disappointment, especially if they believe the 'wrong' interests have been successful in gaining the ear of council representatives. Whatever the reason, the fact that precisely these individuals are most negative is not to be made light of, since they are likely to be quite articulate in making their disappointment known.

Confidence in the Local Administration:
Greatest Among Public Employees and the Satisfied

The last object in our study of local government legitimacy is the local administration. Again we start with an evaluation of confidence. To assess this we employ questions regarding the confidence people have in those employed by local government and an evaluation of the work effort of municipal employees compared with those employed by private business. Responses to these questions are moderately correlated (r = 0.28) and are combined to create an index of 'confidence in the local administration'.

One of the best predictors of confidence in those employed by the municipality is whether or not the respondent is personally employed in the public sector (see column 1 in Table 2.4). This finding is not surprising. People employed by government, be it at the national or local level, are quite reasonably likely to feel as competent, honest and committed to a work ethic as those in private business. The notion that there is a difference is one which is held (and perhaps more actively cultivated?) by those in the private sector. But it is interesting that those with higher education are more skeptical about the local administration, even if the level of education among public employees is on average higher than in the private sector. A similar skepticism is found among the well to do.

Table 2.4 Regression Coefficients for Perceptions Regarding Confidence in the Local Administration and General Performance of Local Government

Independent Variables	Confidence in the Local Administration	Performance of Local Government
Age	0.04	−0.04
Gender (women)	0.02	0.05
Civil status (married/cohabitant)	0.03	0.01
Education	−0.10**	−0.02
Household income	−0.08**	0.05
Occupational sector (public)	0.19***	0.04
Subjective knowledge of local politics	0.11***	−0.01
Subjective local political competence	0.08*	0.26***
Voted for mayor's party (yes)	0.04	0.08**
Scope of municipal activity (acceptable)	0.08**	0.07**
Citizen role orientations		
Consumer	0.07*	0.01
(Tax)payer	−0.17***	−0.14***
Voter	0.01	−0.04
General satisfaction with municipal services	0.20***	0.20***
Size of municipality (population logged)	−0.04	−0.05
Adj. R^2	0.17	0.18
N =	(1,095)	(1,066)

The content of all variables is described in detail in a methodological appendix.
Significance levels: * = 0.05   ** = 0.01   *** = 0.001

By comparison, a sense of familiarity with local affairs, as well as a subjective sense of local political competence, are associated with greater trust and confidence in the local administration. A similar pattern is found for those who find the level of activity of local government to be acceptable rather than too high or too low, and among those who place emphasis on the municipality as a provider of public goods and services (a citizen-consumer role orientation). Not surprisingly, being satisfied with local service provision is quite strongly related to confidence in those employed by local government. At the same time there is nearly as strong a tendency among those who are more concerned with the financing of municipal activity to be more skeptical about the local administration.

For once, we shall also comment on a 'non-finding'. Notice that so far there has always been a significant, mostly negative relationship to municipal size. In this instance this relationship has largely faded away. This suggests that, while the legitimacy of the other political objects investigated shows some correlation with municipality size, this is not the case for confidence in the local administration. From the second column of Table 2.4 we also notice that general evaluations of the performance of local government are not related to municipal size.

Performance of Local Government:
The Subjectively Competent and Satisfied are Most Positive

The last aspect of local government to be analyzed is the way local government performs. According to legal-rational norms, governmental activities should be characterized by open, unbiased and predictable routines. One measure of this property is the 'rigidity' of local government – i.e. whether local government is capable of identifying and reversing erroneous decisions. Our other question concerns 'clientelism' – i.e. whether it makes a difference to have personal connections inside local government, either with politicians or with persons in the local administration. Responses to these questions are moderately correlated (r = 0.27) and are combined into an index relating to the 'performance of local government'.

As can be seen in the second column of Table 2.4, those who display the most favorable perceptions of local government performance are first and foremost those who have a sense of personal local political competence, as well as those who are generally satisfied with municipal services. Much weaker, but still significant, those who voted for the party of the mayor in the most recent election, together with those who find the scope of local government activity acceptable, also evaluate the performance of local government positively. The

The Legitimacy of Local Government 49

latter findings suggest that a partisan and/or 'policy component' may color evaluations of local government performance. This is contrary to norms of legal-rational neutrality, but it is difficult to draw more definitive conclusions on the basis of these findings alone. The only factor that exerts a strong negative impact on perceptions of local government performance is again a citizen role orientation which places an emphasis on municipal finance and spending.

A PRELIMINARY SUMMARY: SATISFACTION WITH MUNICIPAL SERVICES IS THE PRIMARY SOURCE OF LOCAL GOVERNMENT LEGITIMACY

Our findings lead to a number of interesting and not altogether obvious conclusions. These may be summarized in the following points:

- The principal conclusion is that a general sense of satisfaction with municipal services is the primary source of positive perceptions. We suggest this may serve as a foundation for a sense of local government legitimacy. Satisfaction with municipal services runs like a golden thread through the analyses undertaken. Whatever the object of input or conversion, and whatever the property of the object in question, satisfaction with municipal services exhibits one of the strongest positive relationships to favorable assessments.
- The relevance of other factors is more variable. Most consistent after service satisfaction is the relationship for a citizen-taxpayer role orientation toward local government. In this instance, the relationship is negative: those most concerned with municipal finance and spending tend to be more negatively disposed to objects and properties of local government. But with respect to confidence in the municipal council, no significant relationship is evident once the impact of other factors is taken into account.
- A third factor of quite consistent impact, albeit not in a uniform direction or of consistent strength, is the size of the respondent's municipality of residence. In five of the eight analyses undertaken, the relationship is negative. This indicates that local government obtains stronger marks from residents in smaller municipalities, while residents of larger municipalities are less enamored with objects of local self-government. The exception is for views on the indispensability of local political parties, where inhabitants of larger municipalities are more favorably disposed to their existence. When it comes to assessments of the local administration and general governmental performance, size has no significant impact.[31]
- Among the remaining explanatory variables, measures of competence most frequently exhibit a significant independent effect. With only one exception,

this effect is positive. Those with more objective and/or subjective competence tend to perceive properties of the objects more favorably. This applies especially to the perceived performance and responsiveness of objects. One practical implication is that providing more information and expanding opportunities for active citizen involvement appears to entail little risk of weakening positive perceptions of local government and its legitimacy. On the contrary, such efforts could enhance positive evaluations of local government.
- Our two measures of 'policy distance' (viz. whether the respondent voted for the party of the mayor and views on the scope of municipal activity), also bear positively on perceptions relating to local government legitimacy, although in a less systematic fashion.
- Finally, of the assorted social background characteristics considered, only age and education have a somewhat consistent impact. Gender and civil status have no significant independent effect. Household income and occupational sector are relevant only as an exception rather than as a rule. When they are relevant, the effect is in a largely predictable direction. Education has otherwise already been discussed: where relevant, its impact is positive. Somewhat unexpectedly, by comparison, age shows a significant negative relation in four of our eight analyses. This is after effects of education and other variables are taken into consideration. We have already commented upon this finding, suggesting that what may be in evidence here is a generational effect, but this result merits further investigation.

DISCUSSION AND CONCLUSION

These findings offer many prospects for reflection and research. Before concluding, we will briefly take up two obvious questions. First, if satisfaction with municipal services is as critical as would appear in positive assessments of various political objects and properties, what serves to generate satisfaction? To put this another way, how is satisfaction distributed over the population? Are some groups very satisfied while others are not satisfied at all, or is satisfaction more evenly distributed throughout the population?

A preliminary answer to this question is found in Table 2.5. No group identified by the social background variables considered – age, gender, civil status, education, income or sector of employment – is significantly more or less satisfied than others. At least in this sense it would seem that Norwegian municipalities have functioned in an even-handed fashion. To the extent our analysis identifies factors that help account for variation in satisfaction with municipal services (and this is not very great, as an adjusted R^2 of 0.04 testifies), two variables are most

Table 2.5 Regression of Satisfaction with Municipal Services Against Selected Explanatory Factors

Independent Variables	General Satisfaction with Municipal Services
	Beta
Age	0.03
Gender (women)	–0.02
Civil status (married/cohabitant)	0.05
Education	0.04
Household income	0.04
Occupational sector (public)	–0.01
Subjective knowledge of local politics	0.11***
Subjective local political competence	–0.02
Voted for mayor's party (yes)	0.02
Scope of municipal activity (acceptable)	0.15***
Citizen role orientations	
Consumer	0.02
(Tax)payer	0.02
Voter	–0.07*
Size of municipality (population logged)	–0.08**
Adj. R^2	0.04
N =	(1,153)

The content of all variables is described in detail in a methodological appendix.
Significance levels: * = 0.05   ** = 0.01   *** = 0.001

significant. These are a subjective sense of being informed about local affairs and a perception that the present level of municipal activities is as it should be. But the latter variable lies substantively close to general service satisfaction, so it does not provide much explanatory help. We also see that size of municipality and a citizen-voter role orientation are negatively related to general satisfaction with municipal services, although the effects are modest.

The relatively weak relationship between municipal size and general satisfaction is not because size is without general significance. Other research regarding citizen satisfaction with municipal services in Norway shows that municipal size is relevant (Baldersheim et al., 1990; Rose et al., 1994; Indergård 1996; Jensen 1998). But satisfaction varies a good deal depending on the services involved. Residents in smaller municipalities, for example, tend to be more satisfied with services relating to child daycare, care for the elderly and primary schools, whereas residents in larger municipalities are more satisfied with technical and cultural services. This mixed situation contributes to the relatively weak relationship revealed. But even allowing for this, there are obviously other factors that contribute to general satisfaction with municipal services. These factors remain to be identified.[32]

A second question brings us back to our point of departure. That is, what bearing, if any, do our findings have on political participation – voter turnout in particular? Is voting turnout positively or negatively related to perceptions of the different objects and properties of local government considered? Can the recent decline in turnout for local government elections be interpreted as reflecting perceptions of local government legitimacy? A tentative response to this question is found in Table 2.6. Here we display bivariate correlations between the eight indices relating to perceptions of local government and self-reported voter turnout in the 1995 elections. In addition, the bivariate correlation between general satisfaction with municipal services and turnout is reported. The findings are quite clear. With only minor variation, the relationships are uniformly, but weakly positive. In short, there is no strong relationship to be observed. What tendency does exist suggests that those who hold most favorable perceptions are more likely take part in local elections than those who hold less favorable views, and vice versa. This pattern is in keeping with the suggestion that declining participation is an expression of dissatisfaction. But to say on the basis of these findings that the relationship gives evidence of a 'crisis' of local government legitimacy is to exaggerate grossly.

Seen in the light of findings from other research, there is an equally if not more plausible interpretation. This interpretation leads to an alternative

Table 2.6 Pearson Correlation Coefficients for Measures Reflecting Legitimacy of Local Government and Satisfaction with Municipal Services with Voting in the 1995 Municipal Election

Measures	Correlation coefficient	N
1. Confidence in local politicians	0.14***	1,807
2. Performance of local politicians	0.06**	2,265
3. Indispensability of local political parties	0.09***	1,662
4. Performance of local political parties	0.09***	1,690
5. Confidence in the municipal council	0.11***	1,850
6. Responsiveness of the municipal council	0.05*	2,281
7. Confidence in the local administration	0.07**	1,834
8. Performance of local government	0.06**	2,202
9. Satisfaction with municipal services	0.08***	2,401

Correlations based on listwise deletion of cases (for missing values) are generally comparable or of slightly greater absolute value than those displayed here.
Significance levels: * = 0.05   ** = 0.01   *** = 0.001

conclusion; namely that there is no immediate crisis of local government legitimacy. The basis for this conclusion is not only the wellspring of support generated by preponderantly positive assessments of local government service provision, but also the positive perceptions of different objects and properties of local government held by younger age cohorts. Yet these cohorts are known for lower rates of electoral participation, a tendency that has been more pronounced in recent years (cf. Bjørklund, 1999, p. 220). Moreover, other results from the research project this chapter draws upon suggest that the idea that voting is a civic virtue is more weakly endorsed by young people than by older age groups (Rose & Pettersen, 1995, 1997). Low and even declining voting, especially among the young, seems therefore more to reflect a weakening in traditional thinking about voting as a sign of good citizenship than a sign of declining local government legitimacy.

A serious challenge nonetheless remains. If, as our findings suggest, satisfaction with municipal services is a key to positive perceptions of local government, then the critical test is whether or not municipalities can maintain their record of public service delivery. For a large segment of the Norwegian population a citizen-consumer orientation appears to be of primary importance (cf. Rose, 1999). If local governments are not able to meet these expectations, talk of a crisis of local government legitimacy may be more justified. But for the time being, so long as local governments are able to produce services people find useful and of a satisfactory quality, the legitimacy of local government does not seem to be in trouble, despite declining interest in partisan politics and falling local electoral participation.

NOTES

1. Funding for research upon which this chapter draws has been provided by the Norwegian Research Council (*Norges forskningsråd* – NFR, project nr. 103187/510) as part of a broader program of research on local government. The program received contributions from the Ministry of Local Government Affairs, the Norwegian Association of Local Authorities, and the Ministry of the Environment. This support is gratefully acknowledged. The authors alone are responsible for analyses and interpretations presented here.
2. Elections for both bodies are held at the same time, so the marginal cost of casting a ballot in both elections is relatively low once a voter has decided to vote.
3. Direct elections for the county councils were first introduced in 1975. Before then representatives to county councils were indirectly elected by municipal councils.
4. Turnout in the 1963 local elections was in many respects abnormal, since elections took place in the aftermath of a national cabinet crisis that focused and rallied the political interest of many citizens. The general point concerning a long-term decline in local election turnout nonetheless remains the same.

5. In a generic sense this question applies to voting at all levels of government and is at the core of what Lijphart (1997) has referred to as 'democracy's unresolved dilemma'.

6. It should be stressed that this is an altogether different argument than that which accepts, indeed in some cases even finds grounds for consolation in, low turnout due to skepticism over the implications of having high levels of turnout. This latter argument is found in different forms in such works as Berelson (1952), Berelson et al. (1954), Morris Jones (1954), Almond & Verba (1963), Lipset (1963), and, perhaps above all, Schumpeter (1942).

7. Hirschman (1982) has made important observations about shifts in citizen orientations over time and the need to exercise caution in drawing conclusions about relatively short-term shifts.

8. Not all scholars are equally enamored by Weber's work. For a sample of critical remarks, see Schaar (1970), Pitkin (1972), Grafstein (1981) and Beetham (1991). Irrespective of one's view of Weber's work, his treatment of legitimacy represents an unavoidable point of reference.

9. For a general discussion of recent developments in local government in Norway, see Rose (1996).

10. Complete item wording and coding of questionnaire items in the analysis are presented in the appendix to this chapter.

11. The strategy employed here, although indirect and incomplete, is in keeping with a long-standing tradition in the empirical literature. See Listhaug (1995) and Listhaug and Wiberg (1995) for recent examples of work in this tradition. As Listhaug and Wiberg (1995, p.299) remark: 'The concept of legitimacy moves the analytical focus beyond support for political parties, politicians, incumbents, or maybe even institutions. But the analysis of confidence in institutions is at least a step towards measuring legitimacy'. Earlier work includes Litt (1963), Aberbach & Walker (1970), Citrin (1974), Miller (1974), Lipset & Schneider (1983), Listhaug (1984, 1989), Baldassare (1985) and Miller & Listhaug (1990).

12. Some empirical evidence on this condition is offered by Muller and Jukam (1977) for Germany. Work reported in Nye et al. (1997) indicate the same for the U.S., and Listhaug (1984, 1989) and Knutsen (1985) provide similar evidence for Norway. For comparative evidence, see Miller & Listhaug (1990) and several chapters in Norris (1999), especially those by Dalton & Holmberg.

13. The work of Lipset and Schneider (1983) is typical in this regard. The emphasis on trust or confidence in politicians is partly a reflection of survey items included in the early U.S. election studies carried out by the Center for Political Research at the University of Michigan. These studies established a strong research tradition that has subsequently been emulated in many countries.

14. The debate between Citrin (1974) and Miller (1974), however, was one that revolved around the true meaning and content of U.S. election study items – i.e. whether they reflected the legitimacy of specific actors or represented more general measures of regime legitimacy.

15. This summary description is based on findings reported by Aardal & Valen (1995, pp. 200–202) and more recent survey results (cf. Pettersen & Valen, 1995, p.73).

16. Using a broader data base, Dalton (1999, pp. 63–64) takes issue with Listhaug's findings and interpretations.

17. The development of the welfare state in Norway has rightfully been described as the development of 'welfare municipality' (cf. Grønlie 1967, Nagel, 1991). This phrase

The Legitimacy of Local Government

is justified as local governments have occupied a prominent position, as initiators and instruments used by national authorities, for the implementation of welfare programs. For a fuller discussion of local government activities, see Rose (1996).

18. The categorization and review undertaken here to some extent draws on the work of Listhaug (1995) and Listhaug & Wiberg (1995).

19. The government overload literature blossomed in many variants. Some of the better known examples include Crozier et al. (1975), King (1975), Rose & Peters (1978) and Offe (1984).

20. See, for example, Walzer (1989) and Dahrendorf (1994). In the Norwegian context this issue is evident in the work of Eriksen (1993), Eriksen & Weigård (1993) and Hansen (1995).

21. Just what values local government may be intended to serve is a matter of debate. As Sharpe (1970) has noted, autonomy, democracy and effectiveness have been quite prominent historically, but the emphasis placed on these has varied over time and from place to place. For more recent contributions to this discussion, see Chapter Two in Burns et al. (1994), along with King & Stoker (1996) and Pratchett & Wilson (1996).

22. This suggestion is one of several considerations central to the debate over how voters orient themselves and behave in various electoral circumstances. The exchange between Miller et al. (1976) and Popkin et al. (1976) is typical in this respect.

23. Although bearing a close resemblance, this is not necessarily the same as the ritualistic negativism that Citrin (1974) referred to in his debate with Miller. For an interpretation of the Swedish case which draws on similar perspectives, see Holmberg (1981, p.161)

24. The work of Robinson (1976) and Inglehart (1977, 1990) is also relevant in this regard.

25. For an investigation bearing on this issue, see Rose and Pettersen (1999).

26. Knutsen (1985, 1998) and Lafferty (1988) highlight the importance of the public-private occupational dimension in Norway. Andersen (1992) similarly included this variable in his Danish work.

27. At present Norway has 435 municipalities, ranging from under 300 to just over 500,000 inhabitants. The average size is roughly 10,000 residents, but over 55% of all municipalities have fewer than 5,000 inhabitants, and nearly 30% have fewer than 2,500 inhabitants.

28. For a review of postwar developments in the structure of local government, see Rose (1996, pp. 168–172). The most recent public commission report, which was submitted in 1992 (NOU 1992, p.15), suggested that as a rule municipalities should have a minimum of 5,000 inhabitants. This recommendation would eliminate a little more than 55% of all municipalities. The proposal created substantial outcry. The recommendation was toned down in the subsequent White Paper (St.meld. nr. 32 1994–95), but Parliament went further in dealing with this document, by essentially prohibiting mergers of municipalities without local support.

29. In addition to size, we expect conditions relating to 'local political overload' (i.e. strain placed on political actors and institutions in attempting to deal with demands created by local problems) and local living standards to be related to popular perceptions of local political objects. These conditions are investigating in an extension of the findings reported in this chapter.

30. The common rule of thumb that items included in an index should have inter-correlations of 0.30 or greater is based on a reflexive measurement model that presumes

a common underlying 'causal dimension'. A formative measurement model, by comparison, assumes no such dimension and therefore implies no necessary intercorrelation of the items involved. It is the latter measurement model that is most appropriate in this and most of the other indices we have created for the analyses here.

31. These findings serve to underscore the fact that fundamental questions raised by Dahl & Tufte (1973) and Newton (1982) regarding the importance of size for democratic government are by no means definitively resolved, but rather require further elaboration and examination.

32. The work of Mouritzen (1989, 1991) regarding the Danish case is quite interesting in this regard.

REFERENCES

Aardal, B., & Valen, H. (1995). *Konflikt og opinion.* Oslo: NKS-forlaget.

Aberbach, J. D., & Walker, J. L. (1970). Political Trust and Racial Ideology. *American Political Science Review 64*, 1199–1219.

Agger, R., Goldstein, M., & Pearl, S. (1961). Political Cynicism: Measurement and Meaning. *Journal of Politics 23*, 477–506.

Albæk, E., Rose, L., Strömberg, L., & Ståhlberg, K. (1996). *Nordic Local Government: Developmental Trends and Reform Activities in the Postwar Period.* Helsinki: Association of Finnish Local Authorities.

Almond, G. A., & Verba, S. (1963). *The Civic Culture: Political Attitudes and Democracy in Five Nations.* Princeton: Princeton University Press.

Andersen, J. G. (1992). Årsager til mistillid. In: J. G. Andersen, H. J. Nielsen, N. Thomsen, & J. Westerståhl, *Vi og vore politikere,* (161–202). Copenhagen: Spektrum.

Andersen, J. G., Nielsen, H. J., Thomsen, N., & Westerståhl, J. (1992). *Vi og vore politikere.* Copenhagen: Denmark.

Arstein, T. (1966). *Gøy på landet? Bygdelister i norsk lokalpolitikk 1945–1995.* Bergen: University of Bergen. Department of Comparative Politics. M.A. Thesis.

Baldassare, M. (1985). Trust in Local Government. *Social Science Quarterly 66*, 704–712.

Baldersheim, H., Hansen, T., Pettersen, P. A., & Rose, L. (1990). *Publikums syn på kommunepolitikk og kommunale tjenester.* Bergen: Norsk senter for forskning i ledelse, organisasjon og styring (LOS-Senteret). Rapport 90/2.

Beetham, D. (1991). *The Legitimation of Power.* Atlantic Highlands, New Jersey: Humanities Press International.

Berelson, B. R. (1952). Democratic Theory and Public Opinion. *Public Opinion Quarterly 16*, 313–330.

Berelson, B. R., Lazarsfeld, P. F., & McPhee, W. N. (1954). *Voting: A Study of Opinion Formation in a Presidential Campaign.* Chicago: University of Chicago Press.

Berglund, F., & Bjørklund, T. (1994). Kommunevalgene som studieobjekt. Oslo: Institute for Social Research. Paper presented at Nordiske Kommunalforskningskonferanse, 1–4 December 1994, Odense.

Birgersson, B. (1977). The Service Paradox: Citizen Assessment of Urban Services in 36 Swedish Communes. In: V. Ostrom, & F. P. Bish (Eds.), *Comparing Urban Service Delivery System: Structure and Performance,* (243–267). Beverly Hills: Sage.

Bjørklund, T. (1999). *Et lokalvalg i perspektiv.* Oslo: Tano Aschehoug.

Burns, D., Hambleton, R., & Hoggett, P. (1994). *The Politics of Decentralisation: Revitalising Local Democracy.* Basingstoke: Macmillan.

Citrin, J. (1974). Comment: The Political Relevance of Trust in Government. *American Political Science Review* 68, 973–988.
Clark, T. N., & Inglehart, R. (1998). The New Political Culture: Changing Dynamics of Support for the Welfare State and Other Policies in Postindustrial Societies. In: T. N. Clark, & V. Hoffmann-Martinot (Eds.), *The New Political Culture*, (9–72). Boulder: Westview.
Crozier, M., Huntington, S. P., & Watanuki, J. (1975). *The Crisis of Democracy*. New York: New York University Press.
Dahl, R. A., & Tufte, E. R. (1973). *Size and Democracy*. Stanford: Stanford University Press.
Dahrendorf, R. (1994). The Changing Quality of Citizenship. In: B. van Steenbergen (Ed.), *The Condition of Citizenship*, (10–19). London: Sage.
Dalton, R. J. (1999). Political Support in Advanced Industrial Democracies. In: P. Norris (Ed.), *Critical Citizens: Global Support for Democratic Governance*, (57–77). Oxford: Oxford University Press.
Easton, D. (1965). *A Systems Analysis of Political Life*. New York: Wiley.
Eckstein, H. (1966). *Division and Cohesion in Democracy: A Study of Norway*. Princeton: Princeton University Press.
Eriksen, E. O. (1993). *Den offentlige dimensjon*. Oslo: Tano Forlag.
Eriksen, E. O., & Weigård, J. (1993). Fra statsborger til kunde: Kan relasjonen mellom innbyggerne og det offentlige reformuleres på grunnlag av nye roller? *Norsk Statsvitenskapelig Tidsskrift* 9, 111–131.
Gamson, W.A. (1968). *Power and Discontent*. Homewood, Illinois: Dorsey Press.
Gaziel, H. (1982). Urban Policy Outputs: A Proposed Framework for Assessment and Some Empirical Evidence. *Urban Education* 17, 139–155.
Grafstein, R. (1981). The Failure of Weber's Concept of Legitimacy. *Journal of Politics* 43, 456–472.
Graubard, S. R. (Ed.). (1986). *Norden – The Passion for Equality*. Oslo: Norwegian University Press.
Grønlie, T. (1967). Velferdskommune og utjevningsstat, 1945–1970. In: H. E. Næss, E. Hovland, T. Grønlie, H. Baldersheim & R. Danielsen. *Folkestyre i by og bygd: Norske kommuner gjennom 150 år*, (199–281). Oslo: Universitetsforlaget.
Habermas, J. (1994). Citizenship and National Identity. In: B. van Steenbergen (Ed.), *The Condition of Citizenship*, (20–35). London: Sage.
Hansen, T. (1995). Lokalt demokrati ved et vendepunkt? In: T. Hansen, & A. Offerdal (Ed.), *Borgere, tjenesteytere og beslutningstakere*, (75–93). Oslo: Tano.
Harman, H. H. (1967). *Modern Factor Analysis*, 2nd edition. Chicago: University of Chicago Press.
Hirschman, A. O. (1982). *Shifting Involvements*. Princeton: Princeton University Press.
Holmberg, S. (1981). *Svenska Veljare*. Stockholm: Liber Förlag.
Holmberg, S. (1999). Down and Down We Go: Political Trust in Sweden. In: P. Norris (Ed.), *Critical Citizens: Global Support for Democratic Governance*, (103–122). Oxford: Oxford University Press.
Høst, S. (1991). The Norwegian Newspaper System. In: H. Rønning, & L. Lundby (Eds.), *Media and Communication: Readings in Methodology*, (281–301). Oslo: Universitetsforlaget.
Indergård, P. J. (1996). *Misfornøyd eller tilfreds? En analyse av folks tilfredshet med kommunale tjenester*. Trondheim: Norwegian University for Science and Technology. Department of Sociology and Political Science. M.A. Thesis.
Inglehart, R. (1977). *The Silent Revolution: Changing Values and Political Styles among Western Publics*. Princeton: Princeton University Press.
Inglehart, R. (1990). *Cultural Shift in Advanced Industrial Society*. Princeton: Princeton University Press.

Inglehart, R. (1997a). *Modernization and Postmodernization: Cultural, Economic and Political Change in 43 Countries.* Princeton: Princeton University Press.
Inglehart, R. (1997b). The Trend Toward Postmaterialist Values Continues. In: T. N. Clark, & M. Rempel (Eds.), *Citizen Politics in Post-Industrial Societies,* (57–66). Boulder: Westview.
Iyengar, S., & Kinder, D. R. (1987). *News that Matters.* Chicago: University of Chicago Press.
Jensen, R. L. (1998). *Innbyggernes støtte til kommunale tjenester: En kvantitativ undersøkelse.* Trondheim: Norwegian University for Science and Technology. Department of Sociology and Political Science, M.A. Thesis.
King, A. (1975). Overload: Problems of Governing in the 1970s. *Political Studies 23,* 283–296.
King, D., & Stoker, G. (Eds.). (1996). *Rethinking Local Democracy.* Basingstoke: Macmillan.
Knutsen, O. (1985). *Politiske verdier, konfliktlinjer og ideologi. Den norske kulturen i et komparativt perspektiv.* Oslo: University of Oslo Department of Political Science. Ph.D. Dissertation.
Knutsen, O. (1998). *Social Class, Sector Employment and Gender as Political Cleavages in the Scandinavian Countries: A Comparative Longitudinal Study, 1970–95.* Oslo: University of Oslo. Department of Political Science. Research Report 2/1998.
Kornberg, A. (1992). *Citizens and Community: Political Support in a Representative Democracy.* Cambridge: Cambridge University Press.
Lafferty, W. M. (1988). Offentlig-sektorklassen: I støpeskjeen mellom de private og kollektive verdier. In: H. Bogen, & O. Langeland (Eds.), *Offentlig eller privat?,* (139–168). Oslo: FAFO. Rapport 078.
Lane, R. E. (1959). *Political Life: Why and How People Get Involved in Politics.* New York: Free Press.
Lau, R. R. (1982). Negativity in Political Perception. *Political Behaviour 4,* 353–378.
Lau, R. R. (1985). Two Explanations for Negativity Effects in Political Behaviour. *American Journal of Political Science 29,* 119–138.
Lijphart, A. (1997). Unequal Participation: Democracy's Unresolved Dilemma. *American Political Science Review 91,* 1–14.
Lipset, S. M. (1963). *Political Man.* New York: Doubleday.
Lipset, S. M., & Schneider, W. (1983). *The Confidence Gap: Business, Labor and Government in the Public Mind.* New York: Free Press.
Listhaug, O. (1984). Confidence in Institutions: Findings from the Norwegian Values Study. *Acta Sociologica 27,* 111–122.
Listhaug, O. (1989). *Citizens, Parties and Norwegian Electoral Politics.* Trondheim: Tapir.
Listhaug, O. (1995). The Dynamics of Trust in Politicians. In: H-D. Klingemann, & D. Fuchs (Eds.), *Citizens and the State,* (261–297). Oxford: Oxford University Press.
Listhaug, O., & Wiberg, M. (1995). Confidence in Political and Private Institutions. In: H-D. Klingemann, & D. Fuchs (Eds.), *Citizens and the State,* (298–322). Oxford: Oxford University Press.
Litt, E. (1963). Political Cynicism and Political Futility. *Journal of Politics 25,* 312–323.
McDill, E. L., & Ridley, J. C. (1962). Status, Anomie, Political Participation. *American Journal of Sociology 68,* 205–217.
Miller, A. H. (1974). Political Issues and Trust in Government: 1964–1970. *American Political Science Review 68,* 951–972.
Miller, A. H., Goldberg, E. N., & Erbing, L. (1979). Type-Set Politics: Impact of Newspapers on Public Confidence. *American Political Science Review 73,* 67–84.
Miller, A. H., & Listhaug, O. (1990). Political Parties and Confidence in Government: A Comparison of Norway, Sweden and the United States. *British Journal of Political Science 20,* 357–386.

Miller, A. H., Miller, W. E., Raine, A. S., & Brown, T. (1976). A Majority Party in Disarray: Policy Polarization in the 1972 Election. *American Political Science Review 70*, 753–778.
Morris Jones, W. H. (1954). In Defence of Political Apathy. *Political Studies 2*, 25–37.
Mouritzen, P. E. (1989). City Size and Citizens Satisfaction: Two Competing Theories Revisited. *European Journal of Political Research 17*, 661–688.
Mouritzen, P. E. (1991). *Den Politiske Cyklus*. Aarhus: Forlaget Politica.
Muller, E. N., & Jukam, T. O. (1977). On the Meaning of Political Support. *American Political Science Review 86*, 1561–1595.
Nagel, A. H. (Ed.). (1991). *Velferdskommunen*. Bergen: Alma Mater Forlag AS.
Newton, K. (1982). Is Small Really So Beautiful? Is Big Really So Ugly? Effectiveness and Democracy in Local Government. *Political Studies 30*, 190–206.
Nielsen, H. J. (1981). Size and Evaluation of Government: Danish Attitudes Towards Politics at Multiple Levels of Government. *European Journal of Political Research 9*, 47-60.
Nordtun, E. (1992). *Tillit til lokalpolitikarar? Ei undersøking i seks norske kommuner*. Oslo: University of Oslo Department of Political Science, M.A. Thesis.
Norris, P. (Ed.). (1999). *Critical Citizens: Global Support for Democratic Governance*. Oxford: Oxford University Press.
NOU (1992: 15). *Kommune- og fylkesinndelingen i et Norge i forandring*. Oslo: Ministry of Local Government Affairs.
Nye, J. S., Zelikow, P. D., & King, D. C., (Eds.). (1997). *Why People Don't Trust Government*. Cambridge, MA: Harvard University Press.
Offe, C. (1984). *Contradictions of the Welfare State*. London: Hutchinson.
Pettersen, P. A., & Rose, L. (1997). Den norske kommunen: Hva har politikerne ønsket, og hva ønsker folket? In: H. Baldersheim, J. F. Bernt, T. Kleven, & J. Rattsø (Eds.), *Kommunalt selvstyre i velferdsstaten*, (91–126). Oslo: Tano Aschehoug.
Pettersen, P. A., & Valen, H. (1995). Valgkampen. In: A.T. Jensen, & H. Valen (Eds.), *Brussel midt imot*, (67–92). Oslo: Ad Notam Gyldendal.
Pitkin, H. F. (1972). *Wittgenstein and Justice*. Berkeley: University of California Press.
Popkin, S., Gorman, J. W., Phillips, Ch., & Smith, J. A. (1976). What Have You Done for Me Lately? Toward an Investment Theory of Voting, *American Political Science Review 70*, 779–805.
Pratchett, L., & Wilson, D. (Eds.). (1996). *Local Democracy and Local Government*. London: Macmillan.
Rempel, M., & Clark, T. N. (1997). Post-Industrial Politics: A Framework for Interpreting Citizen Politics Since the 1960s. In: T. N. Clark & M. Rempel (Eds.), *Citizen Politics in Post-Industrial Societies*, (9–54). Boulder: Westview.
Robinson, M. J. (1976). Public Affairs Television and the Growth of Political Malaise: The Case of Selling the Pentagon. *American Political Science Review 70*, 409–432.
Rose, L. E. (1996). Norway. In: E. Albæk, L. Rose, L. Strömberg, & K. Ståhlberg, *Nordic Local Government: Developmental Trends and Reform Activities in the Postwar Period*, (159–234). Helsinki: Association of Finnish Local Authorities.
Rose, L. E. (1999). Citizen (Re)orientations in the Welfare State: From Public to Private Citizens? In: J. Bussemaker (Ed.), *Citizenship and Transition in European Welfare States*, (131–148). London: Routledge.
Rose, L. E., & Pettersen, P. A. (1995). Borgerdyder og det lokale selvstyret: Politisk liv og lære blant folk flest. In: T. Hansen, & A. Offerdal (Eds.), *Borgere, tjenesteytere, beslutningstakere*, (36–74). Oslo: Tano Forlag.

Rose, L. E., & Pettersen, P. A. (1997). Civic Virtues and Political Behavior: Theory and Practice in the Norwegian Case. Oslo: University of Oslo. Department of Political Science. Paper presented at the Midwest Political Science Association Annual Meeting, 10–12 April, 1997, Chicago.

Rose, L. E., & Pettersen, P. A. (1999). Confidence in Politicians and Institutions: Comparing National and Local Levels. In: H. M. Narud & T. Aalberg (Eds.), *Challenges to Representative Democracy: Parties, Voters and Public Opinion*, (93–126). Bergen: Fagbokforlaget.

Rose, L. E., & Skare, A. (1996a). *Dokumentasjonsrapport: Undersøkelse om folks forhold til kommunen -- 1993*. Oslo: University of Oslo. Department of Political Science.

Rose, L. E., & Skare, A. (1996b). *Dokumentasjonsrapport: Undersøkelse om folks forhold til kommunen – 1996*. Oslo: University of Oslo. Department of Political Science.

Rose, L. E., Hovland, I.-A., & Skeidsvoll, A. (1994). *Lokal tilhørighet, tilfredshet med offentlige tjenester og holdninger til lokale forhold: Noen resultater fra en landsomfattende undersøkelse, med særskilt vekt på kommunestørrelsens betydning*. Oslo: University of Oslo. Department of Political Science. Working Paper 08/94.

Rose, R., & Peters, G. (1978). *Can Government Go Bankrupt?* New York: Basic Books.

Schaar, J. H. (1970). Legitimacy in the Modern State. In: P. Green, & S. Levinson (Eds.), *Power and Community*, (276–327). New York: Vintage.

Schuman, H., & Gurenberg, B. (1972). Dissatisfaction with City Services: Is Race an Important Factor? In: H. Hahn (Ed.), *People and Politics in Urban Society*, (369–392). Beverly Hills: Sage.

Schumpeter, J. (1942). *Capitalism, Socialism and Democracy*. New York: Harper and Row.

Sharpe, L. J. (1970). Theories and Values of Local Government. *Political Studies 18*, 153–174.

Sharpe L. J. (1988). The Growth and Decentralisation of the Modern Democratic State. *European Journal of Political Research 16*, 365–380.

Statistisk sentralbyrå. (1996). *Kommunestyrevalget 1995*. Oslo/Kongsvinger: Statistisk sentralbyrå. Norges offentlige statistikk C 342.

Statistisk sentralbyrå. (1998). *Stortingsvalget 1997*. Oslo/Kongsvinger: Statistisk sentralbyrå. Norges offentlige statistikk C 478.

St.meld. nr. 32 (1994–95). *Kommune- og fylkesinndelingen*. Oslo: Ministry of Local Government Affairs.

Sørensen, R. (1996). The Legitimacy of Norwegian Local Government: The Impact of Central Government Controls. *Environment and Planning C: Government and Policy 15*, 37–51.

Tucker, L. R. (1971). Relations of Factor Score Estimates to Their Use. *Psychometrika 36*, 427–436.

Walzer, M. (1989). Citizenship. In: T. Ball, J. Farr, & R. L. Hansson (Eds.), *Political Innovation and Conceptual Change*, (211–219). Cambridge: Cambridge University Press.

Weber, M. (1964). *The Theory of Economic and Social Organization*. Translated and edited with an introduction by Talcott Parsons. New York: Free Press.

Wilson, F. G. (1936). The Inactive Electorate and the Social Revolution. *Southwestern Social Science Quarterly 16*, 73–84.

METHODOLOGICAL APPENDIX: OPERATIONALIZATION OF VARIABLES

Explanatory variables

Age
  Continuous variable from 18 to 80 years

Gender
  Men = 0
  Women = 1

Civil status
  Single, widow(er), separated and divorced = 0
  Married and cohabitant = 1

Education
  Primary school = 1
  Secondary school = 2
  University or college = 3

Household income
  Continuous variable with values from under NOK 1,000 to NOK 3,000,000

Occupational sector
  Private sector = 0
  Public sector = 1

Subjective local political knowledge
Based on responses to the following question: 'How well informed do you feel you are regarding that which happens in municipal politics? Would you say that you are very well informed, well informed, somewhat informed, or only slightly informed?'
  Responses were recoded as follows ('Don't know' was treated as missing):
    Only slightly informed = 1
    Somewhat informed = 2
    Well informed = 3
    Very well informed = 4

Subjective local political competence
Based on responses to the following agree-disagree question: '*Local politics* are often so complicated that a person like me can't really understand what it is all about.'

Responses were recoded as follows ('Don't know' was treated as missing):
Agree completely = 1
Agree somewhat = 2
Both agree and disagree = 3
Disagree somewhat = 4
Disagree completely = 5

Voted for party of mayor in 1995 election
Based on responses to following question: 'Can you indicate what party you voted for in the last municipal election?'

Responses were compared with those of the party of the mayor in the same municipality and recoded as follows ('Do not remember' and 'Refused to answer' were treated as missing):
Did not vote or voted for another party = 0
Voted for party of mayor = 1

Scope of municipal activity
Based on responses to following question: 'In general are you in favor of an expansion of municipal activity, do you believe it should be reduced, or is it satisfactory as it is at present?'

Responses were recoded as follows ('Don't know' was treated as missing):
In favor of a reduction / expansion = 0
Satisfactory as it is = 1

Citizen role orientations
The three citizen orientation variables are additive indices based on three questions in which respondents were asked to express how important different aspects of their relations to local government were for them personally. The aspects in question were designed to tap different citizen roles – i.e. being a consumer of public goods and services, a *(tax)payer* financing the provision of public goods and services, and a voter participating in the political arena where public decisions are made. Indices were created by using factor scores from factor analyses of respondent weightings with respect to components in each question relating to the citizen-consumer, citizen-taxpayer and citizen-voter roles, respectively. These indices have a mean of 0 and a standard deviation of 1. Positive values reflect the importance the respondent placed on each role orientation. For more information on these indices, see Rose (1999).

General satisfaction with municipal services
Based on responses to following question: 'Municipalities perform a number of different services. In general, do you think that your municipality's services

are very satisfactory, somewhat satisfactory, only slightly satisfactory, or not satisfactory?'
  Responses were recoded as follows ('Don't know' was treated as missing):
    Not satisfactory = 1
    Less satisfactory = 2
    Neither satisfactory nor unsatisfactory = 3
    Quite satisfactory = 4
    Very satisfactory = 5

Size of municipality
A continuous variable based on population size that has been transformed using a logarithmic transformation (base 10 log).

Dependent variables

All dependent variables used for the analyses reported in Table 2.1 through Table 2.4 are additive indices. They consist of factor scores generated by means of factor analyses of the specific items noted. For a discussion of the use of factor analysis and factor scores in index construction, see Harman (1967) and Tucker (1971). Wording and coding of the items used to create each index are as follows:

General confidence in local politicians
The index is based on two variables, both of which were part of the following battery which was included in a postal questionnaire: 'How much confidence do you have in the following persons, institutions or organizations? Place a checkmark for each of the persons, institutions or organizations mentioned.' The list included both 'Local politicians in the party to which you feel closest' and 'Other local politicians'.
  Responses were coded as follows ('Don't know' was treated as missing):
    Very little confidence = 1
    A little confidence = 2
    Some confidence = 3
    Quite a bit of confidence = 4
    Very much confidence = 5

Performance of local politicians
The index is based on three variables, two of which are the following agree-disagree questions: 'Municipal politicians are too concerned about the next municipal election and too little concerned with the long-term development of the municipality' and 'As a rule one can depend upon the promises of politicians in this municipality.'

Response categories were as follows ('Don't know' was treated as missing and category values were reversed as necessary to assure substantive consistency):
Agree completely = 1
Agree somewhat = 2
Both agree and disagree = 3
Disagree somewhat = 4
Disagree completely = 5

The third variable composing the index is based on responses to the question: 'Do you think that elected representatives in your municipality waste a good deal of our tax money, that they waste some, or that they waste very little of our tax money?'

Responses were coded as follows ('Don't know' was treated as missing):
Waste a good deal = 1
Waste some = 2
Waste very little = 3
Waste no tax money = 4

Indispensability of local political parties
The index is based on the following two agree-disagree questions: 'Democratic government in this municipality would function just as well without political parties' and 'In local elections one should only vote for specific persons, and not for electoral lists as is the case today.' Responses were coded in the same fashion as noted previously.

Performance of local political parties
The index is based on the following two agree-disagree questions: 'The political parties in my municipality are for the most part only interested in people's votes, not in their opinions' and 'The most important problems in this municipality are only to a small degree taken up by local political parties.' Responses were coded in the same fashion as noted previously.

General confidence in the municipal council
The index is based on two variables, both of which were part of the same battery as mentioned with respect to general confidence in local politicians. In this case the following items were used: 'The municipal council in your municipality' and 'The mayor in your municipality'. Responses were coded in the same fashion as noted previously.

Responsiveness of the municipal council
The index is based on three variables, two of which are the following agree-disagree questions: 'Those who sit in the municipal council and make decisions

seldom consider what common citizens think and believe' and 'The municipal council representatives we elect quickly loose touch with common citizens.' Responses were coded in the same fashion as noted previously.

The third variable composing the index is based on responses to the following question: 'How well does the municipal council reflect public opinion in your municipality? Would you say it reflects people's opinions very well, quite well, not very well, or not at all?'

Responses were coded as follows ('Don't know' was treated as missing):
Not at all = 1
Less well = 2
So so = 3
Quite well = 4
Very well = 5

General confidence in the local administration
The index is based on two variables, one which was part of the same battery as mentioned with respect to general confidence in local politicians. In this case the following item was used: 'Municipal employees'. Responses were coded in the same fashion as noted previously.

The other variable was based on responses to the following agree-disagree question: 'Municipal employees work just as effectively as people in private concerns.' In this case the coding of response categories was reversed for substantive consistency.

Performance of local government
The index is based on the following two agree-disagree questions: 'Those who have good personal connections in the municipality are able to have their interests taken care of more easily than others' and 'It is impossible to change a municipal decision, even if the municipality has made a mistake.' Responses were coded in the same fashion as noted previously.

Voting in the 1995 municipal council elections
Based on responses to the following question: 'In the municipal elections last fall nearly 40% of all qualified voters did not vote. What about yourself? Did you vote or not'?

Responses were recoded as follows (all other responses were treated as missing):
Did not vote = 0
Voted = 1

3. CITIZENS, COUNCILORS AND URBAN INSTITUTIONAL REFORM: THE CASE OF THE NETHERLANDS

Bas Denters

The 1990s have witnessed a drop in turnout in Dutch local elections, especially in urban municipalities. This has revived discussion about policies of urban institutional reform. This chapter explores one element of such discussion by examining citizen and councilor perceptions of urban democracy and attitudes towards policy measures to revitalize local democracy. Do the opinions of elected officials on the vitality of local democracy concur with local public opinion? Does the willingness of councilors to adopt institutional reform policies to improve the quality of urban democracy correspond with calls for change among voters? Are there any differences in this type of responsiveness across municipalities, and between political parties? Finally, is it possible to explain individual differences in reformism between councilors: are they simply following their personal ideological preferences or do they respond to their constituents?

These questions will be addressed using citizen surveys in seven urban municipalities (3,000 respondents from Amsterdam, The Hague, Utrecht, Eindhoven, Tilburg, Nijmegen and Zwolle) and a mail questionnaire sent to councilors in these seven cities.[1]

BACKGROUND

The turnout in the Dutch municipal elections of March 1998 witnessed an all-time low. Just 61% of the electorate defied the day's foul weather and went to the polls. In urban municipalities turnout was lower. In cities with over 100,000 inhabitants electoral participation in many cases dropped well below 50%. From an international perspective, such figures may not seem alarming. In the Anglo-American world, for instance, turnout levels in municipal elections tend to be considerably lower (Rallings et al., 1996, p. 64). However, Anglo-American political systems typically employ 'first past the post' (FPTP) electoral systems. Comparative research indicates that the FPTP system, all other things being equal, reduces turnout (Powell, 1980; Jackman, 1987). The Netherlands employs a system of proportional representation (PR) in local elections. For a PR electoral system, turnout in Dutch municipal elections is relatively low. If we compare the Dutch case with other political systems that operate a PR electoral system, the turnout in Dutch local elections is remarkably low (Rallings et al., 1996, p. 64).

One might argue that low electoral participation merely reflects the rather limited powers of Dutch municipalities. Data on the degree of fiscal autonomy of Dutch municipalities appear to corroborate such an explanation. Dutch local government relies heavily on central grants for revenue. In 1995 merely 16% of local revenue was raised through local taxes and levies (Havermans, 1998, p. 90). From an international comparative perspective this is a rather low percentage (Bonnema et al., 1993, p. 34). On the other hand, if we consider the outlays of local government as a percentage of current government outlays, internationally Dutch municipalities are in the middle bracket. In 1997 they spent about 25% of total government outlays in the Netherlands. It is true that in some cases local expenditure is strictly bounded by central regulations but other programs allow for considerable local spending autonomy. Overall, the relative level of local expenditure in the Netherlands is about equal to that in the United Kingdom, and is well above levels in Belgium, Germany, Portugal and France. In all of these latter four countries local government spent less than 15% of the total current government outlays (source: http://www.oecd.org/ puma/stats/window/table8.pd). Interestingly all these four countries also have municipal turnout rates that exceed local electoral participation in the U.K. and the Netherlands (Rallings et al. 1996, p.64). Clearly, the relation between the importance of municipal government and turnout in local elections is not as straightforward as one might initially presume.

In any event, abstention rates of more than 50% are considered by many commentators in the Netherlands to be a potential threat to democratic legitimacy, and even to the right of existence of local government. Historically, this is understandable. After the abolition of compulsory voting in 1970, municipal turnout rates, with occasional ups and downs, initially remained relatively high, fluctuating at around 70%. At the beginning of the 1990s, however, participation declined considerably. When the downward trend first manifested itself in municipal elections in 1990 (with a turnout rate of 62%) panic struck in many a city hall. Numerous municipalities considered reforms to enhance local citizens' political involvement. Most reform measures implied new communication strategies aimed at emphasizing the distinctive features of municipal politics and local parties, as well as stressing the need to improve public information on municipal politics and policies. Other popular reforms related to the internal organization of municipal government (e.g. relations between council, mayor, aldermen and civil servants). These reforms essentially left the balance of power between councilors and the citizenry unaffected. Initiatives aimed at establishing the more direct citizen involvement in local policy-making processes (e.g. neighborhood decentralization, local referenda and initiatives) were much less popular (Gilsing, 1994, p. 24). Parties like *D66* (the progressive liberals) and *GroenLinks* (the Dutch Green Party) do appear to have an impact on the adoption of proposals allowing more direct relations between local government and citizens (Gilsing, 1994, p. 22). In many municipalities, however, local reformers did not succeed in securing sufficient support for their cause.[2]

In this chapter I will endeavor an empirical exploration of the reluctance of municipal councils to reforms their relations with the local electorate. What lies behind this reluctance may not be uniform. On the one hand, reluctance may occur in a setting in which neither citizens nor their representatives see any problem with the functioning of the local political system. As a consequence they do not consider reform necessary. A rather similar situation emerges when citizens and councilors agree that the working of the democratic system may be inadequate but at the same time agree that proposed radical solutions are either inadequate or unacceptable. In both these situations the reluctance to adopt more or less radical democratic reforms occurs in a situation of consensus between local citizens and their representatives. On the other hand, the lack of enthusiasm for democratic reform may take place in a less harmonious setting. Local politicians (1) may not share citizen orientations towards urban democracy and its problems and/or (2) may disagree with popular demands for political change.

For an adequate understanding of the politics of urban democratic reform it is pertinent to determine the settings in which reforms are considered. Therefore relations between public opinion and councilor orientations toward urban democracy and democratic reform will be explored. In this context, I will first compare citizen and councilor orientations towards the working of urban democracy. Subsequently, citizen and councilor attitudes towards democratic institutional reform will be compared. Finally, in order to estimate the impact of constituency attitudes on councilor reform attitudes, I will provide the results of an empirical test of a largely intuitive model of councilor reformism.

CITIZEN AND COUNCILOR PERCEPTIONS OF URBAN DEMOCRACY

One of the factors that might affect the lack of enthusiasm for institutional reform amongst councilors may be that they differ from citizens in their perception of any problems with urban democracy. In this regard, academic and journalistic accounts of the 'crisis of local democracy' identify various problem dimensions. First, it is presumed that *local* democracy is not very salient to people's daily lives. The *nationalization* of the local tax system, the limited fiscal autonomy of municipalities and the increasing number of *national* policies for which municipalities perform merely executive tasks (Hoogerwerf, 1980), are often considered to be indicators of an overwhelming nationalization of local government. Consequently, some claim that Dutch local politics exists in the realm of de-politicized administration (Andeweg & Irwin, 1993, p.163). This should make it difficult to explain the relevance of local politics and local elections to citizens.[3] For this reason, it is often presumed that many citizens lack a genuine interest in local politics.

Second, partly because of the aforementioned factors, people are not well informed about local policies and politics. Local elections are therefore said to be dominated by national rather than local political issues. This nationalization of local elections is often considered to have a negative impact on local electoral accountability. If councilors live under the apprehension that their (mis)conduct in office does not substantially affect their chances at the polls, they may become irresponsible and unresponsive to the local electorate (Newton, 1976, pp. 18–21).

Third, it is argued that local politicians and parties are not well informed about the demands and preferences of the citizenry. This lack of information may be due to various factors. On the one hand, the decline of the traditional Dutch party system is likely to have widened the gap between party politicians and citizens. Traditionally, Dutch political parties were firmly rooted in dense networks of social organizations. These networks were highly segmented, reflecting cleavages in traditional Dutch society by religion (Catholic, Orthodox Calvinists, Dutch Reformed Church) and class (middle versus working class). The organizations that comprised these networks provided for almost every conceivable aspect of life, ranging from trade unions, mass media, schools and universities, sports clubs and other associations for leisure activities. This social system is often referred to as *verzuiling* (or 'pillarization', see Lijphart 1979). From the mid-1960s onwards, this system declined, due to change in the religious and class structure of Dutch society (Irwin & van Holsteyn, 1989, pp. 34–37). Similar developments took place all over Europe (Dalton & Wattenberg, 1993, pp. 198–202). But in the Dutch case the effects may have been particularly upsetting, because of the previous pervasiveness of the 'pillarized' system in all aspects of social and political life. From a political perspective, change resulted in increasing levels of electoral volatility, in dramatic decline in party membership, and in a loosening of previously close relations between parties and organizations within their segment of society. This process of de-alignment has uprooted Dutch parties and deprived them of their natural embeddedness in society. Consequently party politicians may experience increasing problems in keeping in touch with relevant segments of local public opinion. Set against this explanation, with the growth of local government responsibilities, there is evidence that councilors are being absorbed by work inside city halls. This increased workload involves reading policy documents, talking with fellow politicians and civil servants, and preparing for and attending meetings of party groups, policy committees and the council itself. This trend in council work is thought to result in a quasi-bureaucratization of councilors at the expense of their sensitivity to popular preferences (Denters & de Jong, 1992; Denters & van der Kolk, 1993a).

Table 3.1 reflects many aspects of the 'crisis' of Dutch urban democracy. In the seven cities examined here, about three-quarters of citizens consider local government to be relatively unimportant for the conduct of their personal life. Except for Zwolle, a majority of the electorate in each city readily acknowledge that they fail to perceive much difference between local political parties. The table also shows that virtually all the electorate vote for the same party in local and national elections. According to various observers this suggests that local political considerations are not prominent in determining local election outcomes (Newton, 1976; van Tilburg, 1993; van der Kolk, 1997; Depla & Tops, 1998). Finally, Table 3.1 shows that only a minority of local voters is convinced that municipal councils and local parties are well informed about citizen opinions.

In 1991 the members of an inter-university research team (Denters, Depla, Leijenaar, Niemöller, Tops & van Deth) published seven influential case studies on citizens' political views in Amsterdam, The Hague, Utrecht, Eindhoven, Tilburg, Nijmegen and Zwolle. For the first time in the Netherlands, these case studies provided a survey-based description of local political orientations amongst a representative sample of local citizens. At the time this work was first published (Tops et al., 1991), the findings of the case studies caused considerable commotion. Almost a decade later, studies that reach similar conclusions about citizens' political views pass almost without comment. This paper extends the insights provided in this earlier work by continuing to utilize insights on citizen values, but with a primary focus on *councilor perceptions* of citizens and the condition of local democracy in the early 1990s. Taking these views alongside those of citizens allows concurrence between councilor beliefs and

Table 3.1: Councilor and Citizen Orientations on Local Democracy in Seven Dutch Cities

	Amsterdam	The Hague	Utrecht	Eindhoven	Tilburg	Nijmegen	Zwolle	Correlation (tau-b)
Salience								
Councilors	70	84	73	69	74	66	–	
Citizens	28	27	25	22	23	24	–	0.20
Distance	42	57	48	47	51	42	–	(0.29)
Visibility								
Councilors	90	75	77	61	85	85	96	
Citizens	41	33	36	31	33	41	51	0.82**
Distance	49	42	41	30	52	44	45	(0.01)
Interest								
Councilors	33	19	23	21	18	26	11	
Citizens	69	57	69	67	59	64	63	0.49*
Distance	–36	–38	–46	–46	–41	–38	–52	(0.06)
Local factor								
Councilors	32	23	26	24	30	27	30	
Citizens	7	6	10	10	9	6	7	–0.13
Distance	25	17	16	14	21	21	23	(0.34)
Municipality Informed								
Councilors	60	64	32	42	53	35	–	
Citizens	24	19	23	25	27	27	–	–0.28
Distance	36	45	9	17	26	8	–	(0.22)
Parties Informed								
Councilors	68	72	50	72	68	58	–	
Citizens	29	26	30	34	32	31	–	–0.07
Distance	39	46	20	38	36	27	–	(0.42)
Mean Absolute distance	37.8	40.8	30.0	32.0	37.8	30.0	–	

* Kendall's tau-b significant at 0.10 level for a one-tailed test; ** significant at 0.05 level for a one-tailed test.

Question wording: **Salience (councilors)**: How much influence do you think decisions by the municipal government generally have on the daily life of the citizens? *Very much, fairly much,* fairly little, or no influence at all. **Salience (citizens)**: How much influence do decisions by the municipal government generally have on your daily life? Is that *very much, fairly much,* fairly little, or no influence at all. **Visibility (councilors)**: Could you indicate for each statement whether you fully agree with it, agree somewhat with it, *disagree somewhat with it, or fully* *disagree with it?* ... I see hardly any differences between the parties represented in the municipal council. **Visibility (citizens):** Could you tell me for each statement whether you *agree* with it, or disagree with it? ... I see hardly any differences between the parties represented in the municipal council. **Interest (councilors):** Could you indicate how much, you think, citizens in this municipality are interested in local political topics? Most citizens are: not interested, hardly interested, *fairly interested, very interested*. **Interest (citizens):** Are you *very interested* in local political topics here in [name municipality], *fairly interested*, hardly interested or not interested? **Local factor (councilors):** To what extent, do you think, are the results of municipal elections in this municipality determined by local factors (Please indicate a percentage that comes closest to your assessment: 100%–0%). Entries in the table are mean percentage per municipality. **Local factor (citizens):** Determined by comparing answers on questions about voting intention in a local election (if that would take place tomorrow) and about voting intention in a national election (if that would take place tomorrow); Entries in table are percentage of the people providing different answers for these two questions. **Municipality informed (councilors):** Do you think that the municipal government is *sufficiently* or insufficiently informed about what the citizens in this municipality want? **Municipality informed (citizens):** And, do you think that the municipal government of [name municipality] is *sufficiently* or insufficiently informed about what the citizens in this municipality want? **Parties informed (councilors):** Could you indicate for each statement whether you *fully agree with it, agree somewhat* with it, disagree somewhat with it, or fully disagree with it? ... The political parties are well informed about what the citizens in this municipality want? **Parties informed (citizens):** Could you tell me for each statement whether you *agree* with it, or disagree.... The political parties here are well informed about what is going on among the citizens of this municipality. The table gives the percentage of respondents providing the *answers in italics*.

citizen orientations to be assessed. In making this assessment, two indicators are used.

First, *within* each municipality, absolute differences between citizens and councilors are identified through their responses to similar questions about municipal government. If we inspect the results in Table 3.1 it is evident that there is a considerable and consistent gap between citizen orientations towards local politics and councilor beliefs about urban democracy and its citizens. About 70% of councilors feel that their decisions have a considerable effect on citizens' daily lives. Only one-quarter of the citizenry shares this belief. Whereas sometimes the overwhelming majority of councilors claim visible program differences between local parties, most of these distinctions appear to escape popular notice. Likewise, councilors appear to overestimate the 'local factor' in municipal elections. This misperception is not a phenomenon that is restricted to councilors in our seven municipalities. Similar observations were made in a survey of councilors of all municipalities in the Dutch province of Overijsel (Denters & de Jong, 1992, p. 83; van der Kolk, 1997, pp. 141–143) and in Newton's analysis of Birmingham (1976, pp. 14–18). Newton interprets this as an effect of uncertainty: "[t]his uncertainty puts the councilor on his guard and concentrates his attention wonderfully on the things he can influence, or thinks he can influence" (p.19). Finally, local representatives believe that local parties and municipal authorities are relatively well-informed about local public opinion. This sometimes blatantly diverges from popular beliefs. The Hague is a prime example. In this city a two-third majority of the council (the highest percentage in our seven cities) is confident of being well-informed, whereas less than one-fifth of the citizenry agrees with these representatives (the lowest percentage among our seven cases).

The second indicator of agreement (or disagreement) is less exacting, as it focuses on comparison of the aggregate responses *between* municipalities. It asks if there is a positive correlation between the percentage of citizens who agree with a particular questionnaire item and the corresponding percentage of councilors who agree with the same questionnaire item. This measure registers co-variation between aggregate citizen and councilor percentage scores across municipalities. In the final column of Table 3.1 the rank-correlation (Kendall's tau-b) between citizen and councilor percentage scores is reported for the case study municipalities. We have to be cautious with these figures, since these coefficients are based on only six or seven cases.[4] Nevertheless the results provide support for the analysis *within* municipalities, in suggesting no high level of concurrence between aggregate citizen and councilor views *across municipalities*. Only in two instances, for the visibility of party-differences and for citizen interest in local politics, is there a (significant) and relatively strong positive correlation between citizen and councilor views at the municipal level. For the other four explored political views, coefficients are insignificant. Some are even negative.

This lack of agreement may be due to various factors. In some cases, the findings are probably due to inaccurate perceptions on the side of councilors. This appears to be quite likely for the impact of local considerations on electoral choices in municipal elections. In other instances the origins of incongruities may be far less obvious. In the case of perceptions of the saliency of municipal decisions, for example, divergence between citizens and councilors might reflect councilor over-estimation of the importance of their work for ordinary citizens. However, disagreement might just as easily originate from public lack of awareness of the full range of municipal responsibilities, which could lead to misjudgment over the salience of local politics. Similar comments apply to other survey items in Table 3.1.

Either way, disagreement between the views of electors and elected is noteworthy for survey items that may be considered indicators of a healthy local democracy (citizen convictions about the saliency of local issues, public visibility of party differences, a low degree of 'nationalization' of local elections and popular confidence that local parties and authorities are in touch with public opinion). Normatively, disagreement between representatives and their constituents on the condition of urban democracy may be cause for concern among democrats. Empirically, the findings are relevant too. For one, one might argue that an over-optimistic climate amongst councilors is not conducive to councilor support of radical reform of the institutions of urban democracy.

There is one apparent exception to the general pattern of overly optimistic beliefs about citizens and local democracy, viz. councilor beliefs about popular interest in municipal politics. In all our cities, most councilors typically believe that most citizens are at best only scarcely interested in municipal politics (Table 3.2). However, about 60% of citizens in these cities report themselves to be 'fairly' or 'highly' interested in local politics. Even though there is disagreement in absolute terms on this indicator, viewed across the seven municipalities, there is correspondence between citizens' reported interest and councilor perceptions. Municipalities like Amsterdam and Utrecht, which have high levels of self-reported citizen interest, also have relatively high percentages of councilors perceiving popular interest in local politics (see the final column of Table 3.2). The fact remains, however, that there is a considerable absolute distance between citizen and councilor scores. This gap is presumably a manifestation of a more general set of councilor beliefs about the judiciousness of their electorate. In various respects councilors are, rightly or wrongly, not impressed by their constituents' political judiciousness. There are considerable variations between the seven municipalities in this respect. Representatives from Amsterdam and Utrecht are significantly more positive about their electorates than their counterparts in The Hague, Nijmegen and Zwolle. Nevertheless, councilors overwhelmingly think that most citizens do not have an adequate understanding of the complexities of political problems.

Besides, Table 3.2 shows that large councilor majorities consider most constituents as too demanding, with demands that are not oriented to the general public interest. In all seven municipalities, a majority of councilors consider the electorate to be too demanding. However, there are substantial differences between municipalities. Councilors in Amsterdam and Utrecht again hold rather favorable views on their electorate. Even so, about 90% of councilors presume that voters are primarily driven by self-interest. Whether these beliefs are a realistic assessment of the electorate's reasonableness and public regardedness or should be conceived as a display of elitist condescension, is beyond the scope of this paper. However, recent US research does indicate that these views may be on the pessimistic side. Thus Page & Shapiro (1999, pp.112–113) claim that: 'To the question of whether the public is 'rational', in our sense of the word, we remain entirely convinced that the answer is yes [...] In saying that [...] collective policy preferences are 'reasonable', 'responsible' or 'competent', we stand on shakier ground than in discussing rationality. There is more room for disagreement over standards and judgements. We believe, however, that the available evidence tends to support those characterizations as well, and that it also tends to indicate that majorities of the American public are often public spirited and are generally non-tyrannical with respect to minorities'.[5] Yet, rightly

Table 3.2. Councilor Beliefs in Local Political Judiciousness of the Electorate in Seven Cities

	Amsterdam	The Hague	Utrecht	Eindhoven	Tilburg	Nijmegen	Zwolle
Aware of Complexities							
Fully disagree	20	28	19	27	50	54	54
Disagree	33	59	64	54	38	27	35
(Fully) agree	47	12	16	18	12	19	12
Not too Demanding							
Fully disagree	13	28	20	24	18	26	35
Disagree	43	53	50	52	62	59	50
(Fully) agree	43	19	30	24	20	15	15
Oriented to the Public Interest							
Fully disagree	33	50	23	42	32	56	62
Disagree	57	44	64	54	56	44	27
(Fully) agree	10	6	13	3	12	0	12
Positive Evaluation of Citizens	6.47	5.41	6.10	5.52	5.59	5.00	4.92
Scale mean	(N=30)	(N=32)	(N=30)	(N=33)	(N=34)	(N=26)	(N=26)
(ANOVA F-test: F=3.59; sig. = 0.002)							

QUESTION WORDING: **Aware of complexities:** Most citizens are well aware of the complexity of political problems. **Not too demanding** (coding original question reversed): Against their better knowledge, citizens demand too much from politicians. **Oriented to public interest** (coding original question reversed): Most citizens watch their private interests rather than the public interest.

These statements were preceded by the following introduction: Could you indicate whether you strongly agree, agree, disagree or strongly disagree with the statements below? **Positive evaluation of citizen** is a composite measure based on the three items in this table. It is unidimensional (Principal Component Analysis; Kaiser-criterion; first factor explaining 53% of the variance in these items; all factor loadings {more} 0.65) and has a reliability of 0.55 (Cronbach's alfa) which is reasonable for a 3–item scale. The scale ranges from 3 to 12 (the higher the score the more positive the councilor assesses citizens).

or wrongly, the beliefs of Dutch politicians about their voters may provide an additional element to explain the lack of enthusiasm among councilors for more radical democratic reform.

What should nonetheless qualify our understanding of concurrence in assessments of urban democracy between electors and representatives is whether there are differences between political parties (table not included here).[6] Given that councilors themselves perceive differences between parties, it is pertinent to ask whether such purported differences are articulated in disparity in the 'distance' that exists between citizens and councilors for different parties. By breaking down responses into party groupings, we learn that parties of the left (Social-Democrats and Green Left) tend to diverge slightly less from the electorate at large than representatives of the right and the center. Thus the mean absolute distance between councilors and the average score for local citizens per item is 26.8% for the Greens, 27.5% for the Social-Democrats, 35.2%, for the Social Liberals, 36.2% for the Christian-Democrats and 36.0% for the Conservative Liberals.[7] At the municipal level a modest negative correlation is likewise recorded between the dominance of left parties on council and the mean absolute distance between councilors and electorate (Kendall's tau_b = –0.36; N = 6). There may be various interpretations for this relationship. First, it could stem from a relatively greater attentiveness to public opinion on the part of leftist councilors. From such a perspective the relationship results from the role orientations and role behavior of councilors. Second, it may be the intended result or the by-product of the electoral process. It is possible, for instance, that a strongly reform-oriented electorate consciously chooses relatively large numbers of reform-oriented councilors. In this case the correlation is the intended result of 'rational voting'.[8] Alternatively, the urban electorate (that also happens to have reformist attitudes) habitually votes for leftist parties in both local and national elections, and thereby selects predominantly left-wing councilors (who also happen to have reformist attitudes). Here the correlation is an accidental by-product of the electoral process. Third, a strongly reform-oriented local leadership might successfully raise public support for reform policies.[9] A full-fledged analysis to these and other alternative explanations[10] is beyond the scope of this chapter. However, later in this chapter we will discuss some correlates of councilor reformism. This analysis will allow us to assess the relevance of at least some of the factors mentioned here.

In order to put this latter commentary in context, it should be noted that councilors in Social Liberal, Social Democrat and Green Left parties have more favorable views about the electorate than councilors of the Conservative-Liberal Party, in particular (Table 3.3). At the individual level of analysis we find a significant correlation between positive councilor evaluations of the electorate and individual councilor positions on the left-right continuum. This is true irrespective of the method we employ to determine councilor positions on the left-right continuum. If we use rankings based on external judgments of party positions (based on expert judgements, Castles & Mair, 1984, p. 80) or party manifestos (Tops & Dittrich, 1992, pp. 287–289) the rank-correlation is –0.21 (N = 198). If we alternatively

Table 3.3 Beliefs in Local Political Judiciousness of the Electorate for Councilors of Five Major Parties (unweighted across the seven cities)

	Conservative Liberals	Christian Democrats	Social Liberals	Social Democrats	Green Left
Aware of Complexities					
Fully disagree	56	41	33	25	28
Disagree	44	41	39	52	52
(Fully) agree		18	27	23	20
Not too Demanding					
Fully disagree	51	20	18	21	12
Disagree	44	61	46	50	60
(Fully) agree	4	20	36	29	28
Oriented to Public interest					
Fully disagree	59	44	42	33	40
Disagree	37	53	52	57	40
(Fully) agree	4	3	6	10	20
Positive evaluation of citizens	4.41	5.45	5.97	5.87	5.97
Scale mean (ANOVA F-test: F=5.39; sig.=.000)	(N=27)	(N=66)	(N=33)	(N=47)	(N=25)

QUESTION WORDING: see Table 3.2

employ left-right self placements, the correlation is –0.20 (N = 202). Both coefficients are significant at the 0.01 level (using a two-tailed test).

CITIZEN AND COUNCILOR ATTITUDES TOWARDS DEMOCRATIC INSTITUTIONAL REFORM

Political scientists have characterized the Dutch political system as highly resistant to institutional reform (e.g. Andeweg, 1989; van Deth, 1993). One of the reasons for this may be the Dutch political elite's aversion to change. In this section the attitudes of councilors in the seven cities toward democratic reform will be explored. This will involve making comparisons between councilor and citizen opinions. These provide the basis for conclusions on the degree of responsiveness of councilors across municipalities and parties. Table 3.4 presents data on councilor attitudes towards four reform strategies. The first reform strategy implies that local parties should emphasize their ideological and programmatic distinctions. The second measure is the adoption of local opinion polls to improve the council's information on citizen attitudes. The third measure is the introduction of (consultative) local referenda. A referendum, *de facto* if not by law, implies direct citizen influence on local policy-making. The fourth measure is neighborhood decentralization, which implies delegation of authority to (elected) neighborhood councils. Since a referendum, *de facto* if not by law, implies direct influence by citizens on local policy-making, while neighborhood decentralization implies the delegation of authority, it is the third and fourth measures that rank highest in terms of radicalism for the four options considered. This is reflected in Table 3.4. More than 75% of councilors in all municipalities support more distinct programmatic party profiles. Support for the use of opinion polls is less pervasive, but is still considerable. Councilor approval for the use of consultative referenda,[11] and even more so neighborhood decentralization, is consistently much lower. The only deviant case is Amsterdam's broad support for neighborhood decentralization. But the support of Amsterdam councilors is understandable since the Dutch capital adopted a system of intra-municipal decentralization as early as 1981. The adoption of this reform in Amsterdam was predominantly inspired by motives that had little to do with the desire to improve relations between local government and its citizens. More important were specific local factors in the Amsterdam metropolitan region. These revolved around an attempt to improve inter-municipal co-operation in the region by splitting Amsterdam into smaller municipalities, the need to cut municipal expenditure and the desire to reduce problems of controlling the large municipal bureaucracy (ten Berge & van Ruller, 1987).

Overall, the findings show that measures that do not imply major infringements on 'politics as usual' are more popular amongst councilors than referenda and neighborhood decentralization. Unfortunately, as we lack comparable citizen data for the other measures, we are only able to compare the popularity of reform measures amongst citizens and councilors for two of the four items explored above. Table 3.4 clearly shows that popular support for the referendum option is more widespread than citizen approval for stressing party-differences. This is more or less an exact mirror-image of support for these measures among councilors. In truth, levels of concurrence for 'referenda' are relatively low. Although citizen and councilor data for this issue are not strictly comparable, the somewhat scattered evidence in Table 3.4 suggests that agreement may be

Table 3.4 Councilor and Citizen Support for Democratic Reforms in Seven Cities

	Amsterdam	The Hague	Utrecht	Eindhoven	Tilburg	Nijmegan	Zwolle	Correlation (Tau-b)
Profiling parties								
Councilor	80	88	84	97	88	77	–	
Citizens	62	70	65	68	69	66	70	0.41
Distance	18	18	19	29	19	11	–	(0.13)
Opinion polls								
Councilors	77	59	77	76	68	58	–	
Citizens	*73*	90	*86*	–	85	*59*	90	–
Distance	–	–31	–	–	–19	–	–	–
Neighborhood councils								
Councilors	83	26	16	50	24	27	11	
Citizens	–	87	*60*	*61*	84	70	79	–
Distance	–	–61	–	–	–60	–	–68	–
Referenda								
Councilors	52	50	48	56	59	42	41	
Citizens	74	77	*72*	76	80	74	80	0.07
Distance	–22	–27	–	–20	–21	–32	–39	(0.42)

Percentages in italics: Citizen questionnaires in various cities used somewhat different question wording, so results are not strictly comparable. For this reason no distances were computed. For opinion polls and neighborhood decentralization a correlation was not computed, because the number of strictly comparable cases was too small. Because of this a mean average distance ratio per municipality is not reported.
– Figures are not available because the relevant questions were not included in the questionnaire for this city or the question was phrased in such a way that it was not strictly comparable.
Question wording (only for comparable questions): **Profiling parties (councilors):** To what extent would you consider the measures below as desirable to improve the relation between citizens and government? (*highly desirable, desirable,* undesirable, highly undesirable)... (I) Stressing the distinct features of parties. **Profiling parties (citizens):** Could you tell me for each statement whether you *agree with it,* or disagree with it?... Local politics in this municipality would become much more interesting if parties would make their differences clearer. **Opinion polls (councilors):** To what extent would you consider the measures below as desirable to improve the relation between citizens and government? (*highly desirable,* desirable, undesirable, highly undesirable)... (k) regular opinion polls among the citizenry. **Opinion polls (citizens):** Do you find it desirable if the municipal government would conduct opinion polls among the population to find out how citizens think about certain topics (*desirable* or not desirable). **Neighborhood councils (councilors):** To what extent would you consider the measures below as desirable to improve the relation between citizens and government? (*highly desirable, desirable,* undesirable, highly undesirable)... (e) the introduction of neighborhood councils. **Neighborhood councils (citizens):** Could you indicate whether you *fully agree* with these statements, agree, disagree or fully disagree... Citizens should have more say in their neighborhood. **Referenda (councilors):** To what extent would you consider the

even less in the case of neighborhood decentralization. Moreover, Table 3.4 shows considerable inter-municipal differences in support for 'profiling parties' and 'consultative referenda', especially among councilors. The final column of Table 3.4 indicates that the correlation between councilor and citizen views on 'profiling parties' is modest (0.41, although this is not statistically significant, which is no surprise since N=6). The correlation is much lower for 'referenda' (0.07, N=6). Since a high correlation may be interpreted as an indicator for responsiveness, the small size of this coefficient indicates low concurrence between citizen and councilor attitudes towards this more radical democratic reform.

We can extend this point by considering concurrence between electors and representatives by political party (table not included here). For profiling parties, there are only minor variations between the major parties. The only exception arises for Green Left councilors, who are almost unanimously in favor of this reform strategy. They are thereby more out of tune with the electorate than representatives of other parties. In the case of holding referenda, a majority of councilors of leftist parties (Social Liberals 82%, Social Democrats 61%, Green Left 68%) share the electorate's enthusiasm (with 84% support). However, especially among Social Democrat and Green councilors, a sizeable minority oppose referenda. Conservative Liberals and Christian Democrats are even firmer in rejecting this reform (with support levels of only 35% and 20%, respectively).[12] At the cross-municipal level there is no substantial correlation between leftist domination of a council and concurrence in the support of councilors and citizens for these reform measures. The distance between voters and councilors is about as large in cities with a strongly leftist council as it is in municipalities with a less dominant leftist orientation (Kendall's $tau_b = -0.15$ for 'profiling' and 0.00 for 'referenda').

SOME CORRELATES OF COUNCILOR REFORMISM

In this section the origins of councilor attitudes towards democratic reform are explored. The central question is whether councilors are responsive to local public opinion. This analysis is guided by an intuitive model to explain councilor

---

measures below as desirable? (*highly desirable,* desirable, undesirable, highly undesirable) . . . (d) the introduction of a consultative referendum. **Referendum (citizens):** Some people think that the possibilities for citizens to exert influence on the municipal government are too small. Others think that there are already enough possibilities. Would you please indicate what you think of the organization of a so-called referendum, which means that all citizens can vote on the solution of a particular problem. Do you find such a referendum *highly desirable, desirable,* undesirable, or highly undesirable?

attitudes. The model is comprised of four factors. From a populist view of representative democracy, councilors should act as delegates of their constituents. Acting in this way, representatives help realize the populist democratic ideal of an identity between the popular will and governmental decisions (Thomassen, 1991). If councilors adhere to this role orientation we should find a positive association between public opinion and councilor attitudes (Hypothesis 1). Such a relation might also result if councilors, irrespective of their role orientation, were a perfectly representative sample of the population. In this case the representativeness of the assembly will 'automatically' secure congruence between public opinion and councilor attitudes.[13]

Of course studies of elected officials have shown that they are far from a representative sample of the adult population (e.g. Norris, 1996). Moreover, especially in Western European polities, few representatives adhere to a delegate role (e.g. Thomassen, 1991). As party politicians, they are more likely to rely on their political ideology. Therefore, it is presumed that councilor attitudes on particular democratic reforms will be determined by ideological orientations. The predominant ideological dimension in Dutch politics is along a left-right continuum. Leftist positions are typically associated with positive attitudes towards democratic reform (e.g. Laver & Budge, 1992, pp. 26–27). Therefore, the second hypothesis to be investigated is that support for democratic reform meets with more sympathy the more leftist a councilor's ideological orientation. These ideological orientations could also produce concurrence at the municipal level. This may be the intended result or the by-product of the electoral process. As observed previously, a reform-oriented electorate might deliberately vote for reform-oriented councilors. Alternatively, an electorate (that happens to have reformist attitudes) might habitually vote for leftist parties in both local and national elections, and thereby select predominantly left-wing councilors (who happen to have reformist attitudes). Here the correlation is an accidental by-product of the electoral process.

The third hypothesis is that councilor attitudes towards democratic reform will be more positive when they have faith in the judiciousness of the electorate. This effect is particularly likely when the reform proposed implies more direct reliance in policy-making on the public will (as in the case of referenda).

Finally, one might expect that the inclination toward positive evaluations of reform may be higher the more alarming the 'objective' condition of local democracy (Hypothesis 4). In the early 1990s the low level of turnout in municipal elections was considered by many local politicians to be a major indicator of the alleged worrisome condition of urban democracy. Therefore the level of turnout in the 1990 municipal election is included as the fourth explanatory variable.[14]

Table 3.5 reports the results of simple OLS regression analyses of this model for two reforms – stressing party political distinctions and the introduction of a (consultative) referenda.[15] The first conclusion to be drawn from the analysis is that the intuitive model is not very successful in explaining variation in councilor reform attitudes. The amount of variance explained is well below 10% for both dependent variables. Furthermore, if we look at the presumed explanatory variables, in neither case does public opinion have a significant effect on councilor reform attitudes. These results confirm findings at the aggregate level of analysis that citizen preferences regarding democratic reform fail to make an impact on their representatives.[16]

On the other hand, we consistently find that the ideological position of a councilor affects reformist ideas. The more rightist a councilor's ideological position, the less the representative tends to support democratic reform. As indicated previously, ideologically induced preferences may very well be a factor in bringing about a degree of elite-mass concurrence. This may occur (a) as the intended result or (b) as the by-product of the electoral process. For (a), a reformist citizenry might consciously vote for reform-oriented councilors. In this case the correlation is the intended result of 'rational voting'. For (b), a large portion of municipal voters (that per chance have reformist attitudes) traditionally vote for parties of the left. This results in left-wing dominated councils, and consequently a high level of support for democratic reform among local councilors. Here the correlation is an accidental by-product of the electoral process. This second interpretation is no doubt more persuasive, since we know that few voters cast their vote in municipal elections on local political considerations (see Table 3.1).

Table 3.5 Correlates of Councilor Reform Support, Results of OLS Regressions

	Stressing party differences	Consultative referendum
Councilor's left-right placement	–0.14 (0.03)	–0.18 (0.01)
Local public support for a measure	–0.03 (ns)	0.07 (ns)
Councilor faith in public's judiciousness	–0.11 (ns)	0.15 (0.02)
Local turnout in municipal election	–0.19 (0.01)	–0.06 (ns)
R^2 (N)	0.05 (176)	0.07 (201)

Note: Coefficients are standardized regression (beta) coefficients. Figures in brackets refer to the significance level in a one-tailed test.

The effect of the two remaining variables is not consistent. With regard to the first reform measure (stressing party differences), level of turnout affects councilor support negatively. For the more far-reaching reform (referenda), there is no significant effect for electoral turnout. Apparently if councilors are confronted by a relatively low turnout, which is generally considered to pose a challenge to urban democracy, they react by walking the beaten track of representative party democracy rather than opting for (modest) experiments in participatory democracy. The reluctance to opt for referenda is fostered by the low esteem councilors have for the electorate's judiciousness. The regression results show that support for referenda amongst councilors is positively affected by favorable councilor assessments of the electorate's reasonableness.

CONCLUSION

If we consider the attitudes of citizens at the eve of a new millenium, urban democracies in the Netherlands face considerable problems. Citizens do not consider their local government to be particularly important. Neither do they see much difference between local political parties. This results in a decline in turnout rates in municipal elections. Moreover, those who bother to vote overwhelmingly make their electoral choice using criteria that have little or nothing to do with local policies and politics. Last, but not least, only a minority of the local electorate is convinced that their municipal council and local parties are well informed about citizen opinions.

Although many of the city governments investigated were (and still are) considering strategies for the revitalization of urban democracy, this study suggests no widespread councilor support for more radical institutional reforms (e.g. referenda and neighborhood decentralization). This reluctance stems from at least two sources. First, councilors are more positive in their perceptions of the condition of urban democracy than their citizenry. This optimism is not very conducive to councilor support for radical reform of the institutions of urban democracy. Second, councilors disagree with their voters over the desirability of more radical democratic reforms. Even among the most reform-oriented councilors in left-wing parties, support for rather modest proposals, like the introduction of a consultative referendum, is by no means unanimous and lies well below support for this measure amongst the general public. We may safely say that councilor reform attitudes are out of tune with public opinion in their own cities. While councilors prefer improving the system of representative party democracy, electorates seem increasingly to opt for more direct modes of political participation, which largely bypass elections and parties (Klingemann & Fuchs, 1995). Councilor reluctance to support more fundamental democratic reform is in part explained by a strong sense that citizens lack judiciousness in political matters. Such councilor beliefs have important repercussions for their willingness to consider reform. This is borne out by recent attempts to implement forms of 'interactive governance'. These new modes of policy-making are characterized by their openness and by a wide range of opportunities for citizen participation. However, in The Hague as one illustration, recent studies have shown that the willingness of councilors to experiment with a real delegation of authority (e.g. to citizen groups) is negatively affected by their distrust in citizens' reasonableness (Jakschtow, 1998).

As emphasized before, councilor beliefs about voters may be correct, but appearances are against such a position. Political analysts today report that citizens possess "higher cognitive competence," an "increased ability to process complex information," and have "an increase in knowledge about [their] scope for action in the various arena's of society" (Fuchs & Klingemann, 1995, p. 13; also Page & Shapiro, 1999). Perhaps councilors should reconsider Aristotle's famous "theory of the collective wisdom of the multitude." According to Aristotle (1962, p. 123), who was by no means a fanatic democrat, ". . . it is possible that the many, no one of whom taken singly is a good man, may yet taken all together be better than the few, not individually but collectively, in the same way that a feast to which all contribute is better than one given at one man's expense." However, all these arguments assume that councilor reservations against change are genuinely inspired by concern for possible negative effects from reform on the public good. As political scientists, we are professionally tempted to consider a less benign explanation of reluctance to adopt reform, one that is couched in terms of protection of vested interests and power positions.

NOTES

1. For details on data collection and response rates for the citizen survey, see Anker and Hospers (1993). For similar information on the councilor survey see Denters & van der Kolk (1993b). The citizen survey data files can be obtained from the Steinmetz Archive in Amsterdam (e-mail: steinm@swidoc.nl). For information on the availability of the councilor data contact Bas Denters (e-mail s.a.h.denters@bsk.utwente.nl).

2. At the national level, plans to revitalize local democracy through constitutional reform were considered (e.g. the introduction of elected mayors and decisive local referenda). However, proposed national constitutional changes, like many of the more radical local plans, met with considerable opposition and were eventually rejected. Most recently, in 1999, the Dutch Senate rejected a constitutional amendment allowing for decisive local referenda.

3. This nationalization does not necessarily imply centralization and a marginal role for municipal government (Toonen & Hendriks, 1998). One might argue that involvement in

the execution of national policies allows municipalities considerable *de facto* autonomy, and provides municipalities with resources they would not command under a less nationalized system of intergovernmental fiscal relations.

4. These data are based on all seven municipalities, except for the items 'salience', 'municipalities informed' and 'local parties informed'. For these items no councilor data for Zwolle are available.

5. One might object that this optimistic view is not consistent with a high level of nationalization in municipal elections. The argument presented does not imply that voters may not 'rationally' employ information shortcuts as guidance in voting decisions (e.g. national ideologies). Such shortcuts are even likely in the case of electoral behavior and in the light of the relatively limited relevance of local politics to people's everyday life (Popkin & Dimock, 1999; van der Kolk, 1997).

6. I do not provide analysis by municipality of the concurrence of councilors in each party and voters for the same party. In theory this would provide a more satisfactory measure of proximity. In practice, the number of councilors per party group (even for the largest parties on councils) is so small as to render such a detailed analysis almost meaningless.

7. These results are based on all seven municipalities, except for the items 'salience', 'municipalities informed' and 'local parties informed'. For these items no councilor data for Zwolle are available.

8. This is the traditional one-way model of linkage from mass to elite (Hill & Hurley, 1999).

9. This is the model of elite leadership, in which elites are given a prime role in shaping public opinion (Hill & Hurley, 1999).

10. In addition to these models, Hill & Hurley (1999) point to the possibility of reciprocal or interactive relationships between public opinion and elite attitudes and behavior.

11. Support for decisive referenda is even lower.

12. It should be noted, however, that the councilor question used in this analysis refers to consultative referenda. Support for the use of decisive referenda is by no means as broad.

13. I use aggregate support for the two reforms amongst respondents in municipal samples as estimates for municipal public opinion (in Table 4 the values for citizens in the first and fourth rows). These figures are entered as additional contextual variables in the councilor data-file.

14. I use aggregate electoral statistics to determine the turnout for each municipality. These figures are entered as an additional contextual variable in the councilor data-file.

15. For the 'stressing party differences' model I excluded Zwolle because the dependent variable is not available for this municipality. Pairwise deletion of cases is used for missing data.

16. Conversely, local political leaders have failed to persuade the electorate to accept their point of view. The leadership model of public opinion change suggests this persuasion should be effective (see Hill & Hurley, 1999).

REFERENCES

Andeweg, R. B. (1989). Institutional Conservatism in the Netherlands: Proposals and Resistance to Change. In: H. Daalder, & G.A. Irwin (Eds.), *Politics in the Netherlands: How Much Change?*, (42–60). London: Frank Cass.

Andeweg, R. B., & Irwin, G. A. (1993). *Dutch Government and Politics*. Basingstoke: Macmillan.
Anker, H., & Hospers, L. A. (1993). *Local Democracy and Administrative Renewal in Seven Dutch Municipalities*. Steinmetz Archive/SWIDOC.
Aristotle (1962). *Politics*. Harmondsworth: Penguin.
Berge, J. B. J. M. ten, & Ruller, H. van (1987). Binnengemeentelijke Decentralisatie: Amsterdam als innovatie. In: H. A. Brasz, & J. W. van der Dussen (Eds.), *Gemeenten in Verandering: Is er ook Innovatie?*, (16–196). 's-Gravenhage: VUGA.
Bonnema, W., Cuppen, M. H. M., Evers, A. J. M., & Rikken, W. (1993). *Gemeentefinanciën*, Samsom HD Tjeenk Willink: Alphen aan den Rijn.
Castles, F. G., & Mair, P. (1984). Left-Right Policy Scales: Some 'Expert' Judgments. *European Journal of Political Research 12*, 73–88.
Dalton, R. J., & Wattenberg, M. P. (1993). The Not so Simple Act of Voting. In: A. W. Finifter (Ed.), *Political Science: The State of the Discipline II*, (193–218). Washington DC: American Political Science Association.
Denters, S. A. H., & Jong, H. M. de (1992). *Tussen Burger en Bestuur: een Empirisch Onderzoek naar de Positie van het Raadslid in de Overijsselse Gemeenten*. Enschede: CBOO Universiteit Twente.
Denters, S. A. H., & Kolk, H. van der, (Eds.). (1993a). *Leden van de Raad: Hoe Zien Raadsleden uit Zeven Grote Gemeenten het Raadslidmaatschap?*. Delft: Eburon.
Denters, S.A.H., & Kolk, H. van der (1993b). *Local Democracy and Administrative Reform: Opinions of Aldermen and Council members*. Enschede: Universiteit Twente Faculteit der Bestuurskunde.
Depla, P., & Tops, P. W. (1998). De Lokale Component bij Raadsverkiezingen: de Invloed van de Gemeentegrootte. In: S. A. H. Denters, & P. A. Th. M. Geurts (Eds.), *Lokale Democratie in Nederland: Burgers en hun Gemeentebestuur*, (141–157). Bussum: Coutinho.
Deth, J. W. van 1993). De Permanente Crises in de Nederlandse Politiek. *Acta Politica 28*, 251–272.
Fuchs, D. and Klingemann, H. D. (1995). Citizens and the State: A Changing Relationship?. In: H. D. Klingemann, & D. Fuchs (Eds.), *Citizens and the State*, (1–23). Oxford: Oxford University Press.
Gilsing, R. (1994). Lokale Bestuurlijke Vernieuwing in Nederland. *Acta Politica 29*, 3–36.
Havermans, A. J. E. (1998). Gemeentelijke Taken en Gemeentefinancien. In: A. F. A. Korsten, & P. W. Tops (Eds), *Lokaal Bestuur in Nederland: Inleiding in de Gemeentekunde*, (77–103). Alphen aan den Rijn: Samsom.
Hill, K. Q., & Hurley, P. A. (1999). Dyadic Representation Reappraised. *American Journal of Political Science 43*, 109–137.
Hoogerwerf, A. (1980). Relaties Tussen Centrale en Lokale Overheden in Nederland. *Beleid en Maatschappij 6*, 330–346.
Irwin, G. A., & Holsteyn, J. J. M. van. (1989). Decline of the Structured Model of Electoral Competition. In: H. Daalder & G. A. Irwin (Eds.), *Politics in the Netherlands: How Much Change?* (21–41). London: Frank Cass.
Jackman, R. W. (1987). Political Institutions and Voter Turnout in the Industrial Democracies. *American Political Science Review 81*, 405–423.
Jakschtow, K. (1998). *Het Poldermodel in een Grote Stad: een Onderzoek naar de Opvattingen van de Haagse Gemeenteraadsleden over de Ontwikkeling van Samenspraak*. Enschede: Faculteit der Bestuurskunde UT (afstudeerscriptie).
Klingemann, H. D., & Fuchs, D. (Eds.). (1995). *Citizens and the State*. Oxford: Oxford University Press.

Kolk, H. van der (1997). *Electorale Controle: Lokale Verkiezingen en Responsiviteit van Politici.* Enschede: Twente University Press.

Laver, M. J., & Budge, I. (1992). Measuring Policy Distances and Modelling Coalition Formation. In: M. J. Laver, & I. Budge (Eds.), *Party Policy and Government Coalitions,* (15–40). Basingstoke: Macmillan.

Lijphart, A. (1979). *Verzuiling, Pacificatie en Kentering in de Nederlandse Politiek.* Amsterdam: De Bussy.

Newton, K. (1976). *Second City Politics: Democratic Processes and Decision-making in Birmingham.* Oxford: Clarendon.

Norris, P. (1996). Legislative Recruitment. In: L. Leduc, R. G. Niemi, & P. Norris (Eds.), *Comparing Democracies: Elections and Voting in a Global Perspective,* (184–215). Thousand Oaks: Sage.

Page, B. I., & Shapiro, R. Y. (1999). The Rational Public and Beyond. In: S. L. Elkin, & K. E. Soltan (Eds.), *Citizen Competence and Democratic Institutions,* (93–116). University Park: Pennsylvania State University Press.

Popkin, S. L., & Dimock, M. A. (1999). Political Knowledge and Citizen Competence. In: S. L. Elkin, & K. E. Soltan (Eds.), *Citizen Competence and Democratic Institutions,* (117–146). University Park: Pennsylvania State University Press.

Powell, G. B. (1980). Voting Turnout in Thirty Democracies. In: R. Rose (Ed.), *Electoral Participation,* (5–34). Beverly Hills: Sage.

Rallings, C., Temple, M., & Thrasher, M. (1996). Participation in Local Elections. In: L. Pratchett, & D. Wilson (Eds.), *Local Democracy and Local Government,* (62–83). Basingstoke: Macmillan.

Thomassen, J. J. A. (1991). Empirical Research into Political Representation: A Critical Reappraisal. In: H. D. Klingemann, R. Stöss, & B. Wessels (Eds.), *Politische Klasse und politische Institutionen: Probleme und Perspektive der Elitenforschung,* (259–274). Opladen: Westdeutscher Verlag.

Tilburg, M. F. J. van (1993). *Lokaal of Nationaal?: Het lokale karakter van de Gemeenteraadsverkiezingen in Nederlandse GFemeenten 1974–1990.* 's-Gravenhage: VNG Uitgeverij.

Toonen, Th. A. J., & Hendriks, F. (1998). Gemeenten en Hogere Overheden. In: A. F. A. Korsten, & P. W. Tops (Eds.), *Lokaal Bestuur in Nederland: Inleiding in de Gemeentekunde,* (122–134). Alphen aan den Rijn: Samsom.

Tops, P., & Dittrich, K. (1992). The Role of Policy in Dutch Coalition Building, 1946–81. In: M. J. Laver, & I. Budge (Eds.), *Party Policy and Government Coalitions,* (277–311). Basingstoke: Macmillan.

Tops, P. W., Denters, S. A. H., Depla, P., van Deth, J. W., Leijenaar, M. H., & Niemöller, B. (1991). *Lokale Democratie en Bestuurlijke Vernieuwing in Amsterdam, Den Haag, Utrecht, Eindhoven, Tilburg, Nijmegen en Zwolle (Zeven Delen).* Delft: Eburon.

4. MUNICIPAL RESPONSIVENESS TO LOCAL INTEREST GROUPS: A CROSS-NATIONAL STUDY

Keith Hoggart

In an influential publication on democratic practice, Putnam (1994) argued that the strength of civic communities in Italy plays a fundamental part in securing a less clientelist local politics, greater public confidence in governmental institutions and strong, responsive and effective representative public institutions. Fundamental to Putnam's case is an association of strong civic communities with key characteristics of local populations. In the Italian case, Putnam draws attention to newspaper readership and the existence of cultural associations as two central contributors to this cause, although he also identifies elements in local electoral systems as indices of democratic strength. That Putnam is convincing on this latter point largely stems from the historical analysis he undertakes, so the conditions under which electoral systems were introduced are effectively contextualized. Providing such a temporal framework for interpretation is critical, if organizational form is assumed to lead to uneven democratic practice. Reform government initiatives in the USA offer a sufficient reminder that organizational forms promoted in the 'cause of democracy' can anticipate uneven group advantages (e.g. Hays, 1964; Weinstein, 1968). Assumptions about associations between city-wide social attributes and patterns of political action also require careful treatment, as seen in rather naive assumptions linking ethnicity and working class populations to self-interested, private-regarding behavior (e.g. Hahn, 1970). Such simplistic linkages fail to recognize that self-interested actions can take a variety of forms, and are likely to vary with local circumstances (Swanstrom, 1985). Precisely this understanding lies behind the claims many commentators from the political left make about the actions of interest groups. They are seen to distort local policy-making, as city governments are held to respond more positively to the so-called privileged groups that represent businesses, taxpayers and the middle/upper social classes. Concomitant with this, a less welcoming embrace is envisaged for the proposals of minority and lower income groups (e.g. Chekki & Toews, 1987; Elkin, 1987).

The rationale for this view comes not simply from theoretical perspectives on societal organization but also from empirical evidence on the activities of interest groups. Exemplifying the direct link between theory and expectations of support for advantaged interests, Stone (1980, p. 979) explains that: "... because officials operate within a stratified system, they find themselves rewarded for cooperating with upper-strata interests and unrewarded or even penalized for cooperating with lower strata interests" (also Sjoberg et al. 1966). Adding to their ease in securing desired goals, analysts have argued that wealthier groups also find it easier to unite in concerted action, for the sharing of common interests amongst the upper-strata is stronger than amongst the more socially fragmented lower-strata (Urry, 1986). Although subject to considerable debate and controversy at the local level since Hunter's (1953) thought provoking study of Atlanta, a great deal of research has revealed that closely integrated social networks can be traced at the national level amongst corporate and upper class interests (e.g. Domhoff, 1975; Stokman et al. 1985). Whether or not such relationships can be regarded as influential at the local level is open to question. While there are countries in which analysts see close integration of the national and the local (see Duncan & Goodwin, 1988), in other cases ties are weak (e.g. Page & Goldsmith, 1987), perhaps out of necessity given legal requirements of federalism. Yet the case that is put for governmental bias in favor of advantaged groups is not simply dependent on innate elements in the structure of society. For one there is widespread evidence, from a large number of countries, that the composition of organized interest groups predominantly reveals a strong bias toward the wealthy and the better educated (e.g. Verba & Nie, 1972). In addition, analysts point to a wide range of investigations that provide empirical evidence illustrating the limited effectiveness of less advantaged interest groups. Lipsky (1968) offers a clear picture of this, in highlighting how difficult it is to sustain support amongst less advantaged groups when they seek to produce policy change. Not unexpectedly, given the dire consequences of so many public policy decisions for less advantaged groups (e.g. Fried, 1966),[1] alongside a long history of limited influence on governing bodies, such groups find it difficult to maintain a coherent front against outside threats for long. In this context, those in established positions of authority can commonly weaken the lobbying efforts of disadvantaged groups simply through delay (Dearlove, 1974). As Lipsky (1968) reminds us, what becomes critical for effective pressure to be exerted in such circumstances is appeal to a broader audience.[2] But the capacity to articulate a case in a manner that draws in outside support is unevenly distributed. Potentially significant in this regard is Bondi's (1988) research, which shows that the success of neighborhood opposition to local government initiatives is not tied principally to the socio-economic composition of local populations but to whether groups have access to critical resources that can be used against municipal initiatives. Groups comprised mainly of those with professional and managerial occupations are more likely to possess critical resources, like access to the media, or legal and accounting expertise, but even groups that are dominated by manual workers can secure some of these resources. One example is in residential neighborhoods experiencing the early stages of middle class gentrification, which introduces resources into a predominantly low-income neighborhood, that can be drawn on if the neighborhood is 'threatened'. Yet we cannot assume that the more likely availability of professional expertise amongst the middle classes automatically leads to effective action. King (1979) makes this point clear in reviewing 1970s middle class protest groups in the U.K.; finding that many of them were politically unsophisticated. This point has been re-emphasized recently by the 'reliance' of middle class groups on (radical) 'eco-warriors' to sustain their opposition to developments that threaten their neighborhoods (e.g. Doherty, 1998; Griggs et al., 1998). An interesting contrast arises between King's examination and the actions of so-called environmental protests by these middle class neighborhood groups. In the King (1979) paper the protests that were of concern were those related to taxation and spending – fiscal populism in effect. As Clark's work has long shown, local (and national) politicians have captured these sentiments, so in many places they are now integrated into policy-making processes (Clark & Ferguson, 1983). Piven & Cloward (1979) make a similar point about poor people's movements. They warn us that when poor people's movements develop messages of broad appeal, or effectively challenge established political practices, they are commonly drawn into established power hierarchies, with future action mediated through the bureaucracies whose operational rules militate against the poor (Sjoberg et al. 1966). One indicator of this kind of integration is seen in official environmental agencies in the U.K., which are encouraging middle class rural residents to report environmental infringements by farmers (Lowe et al., 1997). What started as in-migrant middle class complaints about farm practices (Newby, 1980) has been incorporated in public agency strategy, with the protests and conflict engendered being between fractions of the middle classes.

There are two points to take from this. The first is simply that the protagonists in interest group conflicts do not line up in ways that are as straightforward as bourgeois – proletariat class conflict suggests. The second is that the nature of interest group politics changes over time, as particular issues are dampened down by incorporation into mainstream political practices, and as new controversies arise. Critical to understanding interest group activity is an appreciation of how such change occurs. The protests of 1968 provided a very visible message that the relationship between governments and their citizens were changing (Caute, 1988), with resistance to government initiatives occurring in all continents, with marked intensity. Perhaps 1968 might be seen as a unique event, as many disruptions calmed down quickly after that year. But 1968 was the tip of the iceberg, for what came to the surface in that year represented an underlying questioning of governmental actions. This questioning not only occurred across a broad canvass but also revealed that grievance over the direction of societal change was more complex than division based on social class or social inequalities allied to race/ethnicity. The 1960s riots in U.S. cities might be interpreted as a classic dissonance between the interests of the advantaged and the disadvantaged (Button, 1978), but this interpretation does not sit comfortably with the rise of the environmental movement (Lowe & Goyder, 1983). The longstanding women's camp at Greenham Common, which expressed opposition to the U.S. nuclear presence in Britain, alongside recent protests in the U.K. to retain traditional countryside activities (like fox hunting) in the face of liberal, animal-rights campaigners (Hart-Davis, 1997; George, 1999), both highlight the growing incidence of interest groups being comprised of mixed social classes, as well as those in similar class positions taking opposing positions.

As Clark & Inglehart, amongst others, have argued, the social values that dominate today's advanced economies are less easily captured by social class distinctions (Abramson & Inglehart, 1995; Clark & Hoffmann-Martinot, 1998). Lines of division between interest groups are increasingly likely to have a cultural or social dimension, as opposed to an economic one. Moreover, shifts in prevailing societal values have seen growing economic and social individualism, so it has become more difficult to predict value dispositions from individual or household socio-economic circumstances (Clark & Hoffmann-Martinot, 1998, p. 11). In this context it is worth noting how, even in countries where electoral politics has long been cast along lines of social class differentiation, the explanatory power of social class in voting choices has declined sharply (e.g. Heath et al., 1985). Add to this a growing disenchantment with hierarchies in politics, with citizens seeking to promote specific issues, as opposed to accepting that "the government will act as a neutral arbiter" (Clark & Hoffmann-Martinot, 1998, pp. 12–13). Even if elected leaders wish to respond to particular issues, shifting alliances amongst the population make for greater uncertainty in policy-making, especially if politicians have an eye for re-election. Yet the extent to which core values have shifted still requires more empirical verification. Evidence on changing values certainly exists (e.g. Abramson & Inglehart, 1995), but as Clark (1994b, 1996) points out, when it comes to public policy corresponding with value change the link is less direct. It appears that the political 'reality' of intervening circumstances (including resource availability) has to be taken into account in linking citizen values to public policies. Some leaders appear to be less persuaded by the new values of their electorate than others. In a context in which little research has as yet been completed on new value stances and their policy consequences, it is necessary to chart adoption and resistance to the New Political Culture (NPC) by exploring political systems around the world (Clark & Hoffmann-Martinot, 1998). As political histories generate distinctive political cultures, this could feed into the selective policy adoption of values associated with the New Political Culture, and opposition to others.

The aim of this chapter is to offer an exploration of one element of NPC ideas, by examining differential responsiveness to interest group spending preferences in seven countries. The chapter draws inspiration from an earlier paper by Hajnal & Clark (1998), which examined responsiveness to interest groups in U.S. cities. This chapter offers a narrower database than that paper,[3] but extends its focus by looking at more than one country. The central issue in this chapter is an exploration of cross-national differences in municipal responsiveness to interest groups. The approach taken in exploring this issue is to ask what factors are associated with inter-municipal differences in responsiveness to interest groups. The groups that are explored are those of business, taxpayers, low-income residents and minorities. These are taken to represent different positions in the much-touted divide between advantaged (business, taxpayer) and disadvantaged groups (low-income, minority). In addition, neighborhood organizations are explored, as a interest lobby that does not sit comfortably inside traditional class politics divisions, but which provides a point of reference for issue-based politics, which NPC adherents might favor. The specific questions raised in the chapter are how far we can explain inter-municipal differences in responsiveness using measures that represent a class-centered politics and the New Political Culture. This leads to a consideration of whether these factors diverge across nations in a manner that draws out fundamental differences between them. The basis on which cross-national politico-cultural differences lead to dissimilarities in political action has long been the subject of commentary in political science (e.g. Verba, 1987). Heidenheimer & associates (1990) provide one illustration, in their commentary of characteristic national styles.

In the U.K., for example, they see processes of extensive consultation, avoidance of radical policy change and reluctance to take action against well-entrenched interests (note for example the often stated claim that local government in the U.K. is reluctant to be receptive to interest group representations; e.g. Hampton, 1987).[4] By contrast, the French style is said to show a greater preparedness to undertake radical policy shifts, even against the resistance of strong vested interests, while the Swedish approach is to accept radical change after extensive consultation, which might require wearing-down and converting opposing groups (Heidenheimer et al., 1990, p. 350). How far such distinctions can be identified in cross-national disparities in factors associated with uneven responsiveness to local interest groups is the primary question for this chapter.

CROSS-NATIONAL DIFFERENCES IN RESPONSIVENESS BY MUNICIPALITIES

As with the research undertaken by Hajnal & Clark (1998), the data base on which this chapter is drawn is derived from surveys in which local leaders rate their own municipality's responsiveness to demands placed on them by local interest groups. The FAUI database is particularly apt in this regard, as it provides a cross-national survey that uses a common array of questions, alongside appropriate political and socio-economic measures of the framework in which municipal governments reach decisions. This reliance on municipal leader reports is somewhat problematical, if not approached from a critical frame of reference. For one, the FAUI data elicits municipal leader views, but is unable to compare these with the views of local interest groups themselves. In truth it would be difficult to imagine how a sample of such groups could be found that would make for valid cross-national comparisons, except perhaps in very narrow policy fields.[5]

At least in terms of local leader evaluations of municipal responsiveness to interest groups, there are clear disparities between the seven nations examined in this study (Table 4.1; Figure 1.3).[6] In examining these differences, this chapter does not focus on cross-national differences as such. Without a series of other measures to help evaluate the validity of the measures used here, focusing simply on cross-national divergence is unwarranted. For sure, the generally low scores for the United Kingdom seem consistent with reports in the political science literature about a lack of responsiveness by British political leaders to pressure groups (e.g. Heidenheimer et al., 1990). Yet the consistently high responsiveness scores for Germany and Japan perhaps suggest that caution is merited in making cross-national judgements (e.g. compare the values here with

Table 4.1 Mean Average Score for Favorable Responses to Interest Groups

Group type	Australia	Canada	France	Germany	Japan	UK	USA
Business groups	33.8	41.4	32.2	53.0	56.8	29.6	41.7
Taxpayer groups	17.8	28.4	16.5	34.8	45.7	*	22.2
Neighborhood groups	34.8	34.9	46.1	81.2	54.1	31.9	37.8
Low-income groups	31.5	29.3	*	46.7	47.3	37.8	28.1
Minority groups	24.4	22.1	*	40.9	*	27.1	26.9

* No data, as this interest group was not included in the survey for this nation.
The higher the value recorded in this table, the more responsive municipal leaders are to these interest groups. The scale for these scores ranges from 0 to 100, with zero for interest groups municipal leaders whose spending preferences are almost never responded to, and 100 for those that are responded to almost all the time.

the broader perspective on who influences policy-making in Japan and the USA that is offered by Kobayashi in Chapter Ten). Municipal leader responses were also conditioned by the cultural setting in which they operated. Hence, there might be cross-national differences in interpretations of what being responsive to local interest groups means. For those of a cynical mind, the possibility of mutual disturbance in responses to survey items should also be considered. Thus, in a setting in which politicians hold that they operate in a democratic way, might there not be some linkage between reports on how active interest groups are and how responsive municipal governments are to them? This possibility should not be downplayed, yet its potency for this investigation is not strong. This is suggested by three factors. First, using more than one data source, Hajnal & Clark (1998) showed that in the United States activity rates for interest groups had a strong bearing on municipal responsiveness to them. Second, when this relationship is explored cross-nationally, the link is not found to be universally strong (Table 4.2). Indeed, in Germany, municipal leaders report no relationships between neighborhood association activity levels and municipal responses to them. Third, in exploring responsiveness in this chapter, analysis does not simply examine local leader evaluations, but additionally explores responsiveness in the context of group activity rates. In this way, the relative efficiency of interest group actions can be compared; for when they assessed municipal responsiveness, leaders also evaluated interest group activity rates. The juxtaposition of these two assessments in the same questionnaire enables researchers to pinpoint cases where municipal leaders think groups are active but ineffective, as well as the converse.

Starting with an examination of municipal responsiveness scores alone, the analysis undertaken here separates municipalities into those in which groups

Table 4.2 Pearson Correlations for Activity Level and Responsiveness to Interest Groups

Group type	Australia	Canada	France	Germany	Japan	UK	USA
Business groups	0.61	0.30	0.77	0.33	0.56	0.42	0.61
Taxpayer groups	0.46	0.52	0.70	0.35	0.40	*	0.67
Neighborhood groups	0.63	0.38	0.79	-0.06	0.49	0.40	0.66
Low-income groups	0.52	0.40	*	0.32	0.59	0.38	0.59
Minority groups	0.59	0.59	*	0.17	*	0.64	0.68

* No data. These coefficients use the raw scores that municipal leaders gave to each interest group for activity (no activity – the most active of all) and responsiveness scores (almost never responded to – responded to almost all the time).

are responded to more positively than their respective national norms. The reason for focusing on this binary divide is twofold. First, because assessing responsiveness scores relative to national mean averages lessens the chance of distortion resulting from uneven cross-national interpretations of 'responsiveness'. Associated with this, as the survey item on which the analysis is based focused on responsiveness to spending preferences, it is feasible that dissimilarities exist in the financial demands made on local governments in different nations. This could result from dissimilar legal or cultural specificity, either of which strengthens the case for analyzing inter-municipal variations within rather than between nations. Secondly, because interest group activity is inevitably multidimensional, as are municipal responses to them, it is advisable to be cautious in expecting leader evaluations to provide responsiveness scores that are equivalent to interval scale measurement. This point is similar to that for cross-national differences, only here the issue is more about categories in an evaluation scale. The issue is, how far is it feasible to conceptualize leader responses to a question on whether they respond to interest group spending preferences 'about half of the time' or 'less that half of the time' as interval scale measurements? In a multidimensional and dynamic political setting, it was decided to focus on broad categories rather than treating the scores leaders provided as interval scale.

Most often evaluation of propositions in this chapter is undertaken by comparing places whose responsiveness to interest groups falls above and below their respective national mean average. The raw (five or seven point) scales on which leaders rated their responsiveness were used to check these category-based tests. These checks indicate that the categories identify stronger associations between responsiveness and measures of both municipal governance and local socio-economic conditions. This is illustrated in Table 4.3 and Table

4.4. The first table gives the results of analysis of variance tests comparing municipalities with above/below average responsiveness rates for business groups. The second table presents Pearson's correlations for raw scale measures with the same governance and socio-economic variables. The variables in these tables are used throughout this paper. Their measurement and analytical significance is explained in the Appendix to this chapter, alongside principal propositions from NPC theorizing.

What is most striking about municipal responsiveness to business groups is how weak relationships are with underlying governance and socio-economic conditions. Most obviously, as Hajnal & Clark (1998) report for the USA, there is a strong relationship between perceived levels of business group activity and municipal responses to the same. This is a feature that exists for all of the

Table 4.3 Municipal Responsiveness to Business Groups – Above/Below National Mean

Variables	Australia	Canada	France	Germany	Japan	UK	USA
Governance factors							
Parties more on political right				−			
Leaders favor more public spending							
Favor redistributive spending						+	
Socially conservative leaders				*			+
Strong local party organizations	+				+	+	+
% female councilors			*	−			−
Prepared to break rules to help people	+	+	*	*			
Prepared to take unpopular decisions				*			+
Importance of media	+	+	*	*			
General interest group activity	+	+	+		+	+	+
Socio-economic factors							
Higher social inequality			−	*		*	
% foreign residents				*			
Mean education levels			−	*		*	
Mean income levels			+	*		*	
% blue collar workers			*	*	−		
% professional workers			*	+	*		
Youthful population			−	*			

* No data. This table only shows relationships that are 0.10 statistically significant in analysis of variance tests (as 'Prepared to break rules to help people' is a binary variable, chi-squared was used). The variables in the left-hand column are explained in the Appendix.
+ Here places with above the national mean average responsiveness to this interest group record higher scores for the variable in the left-hand column than places below the national mean average.
− Here places with above the national mean average responsiveness to this interest group record lower scores for the variable in the left-hand column than places below the national mean average.

Table 4.4 Municipal Responsiveness to Business Groups – Pearson Correlations

Variables	Australia	Canada	France	Germany	Japan	UK	USA
Governance factors							
Parties more on political right				−0.259			
Leaders favor more public spending							
Favor redistributive spending							0.114
Socially conservative leaders					*		
Strong local party organizations	0.270				0.523	0.345	0.219
% female councilors			*	−0.415			
Prepared to take unpopular decisions					*		
Importance of media	0.272		*	*			
General interest group activity	0.602	0.699	0.376		0.596	0.277	0.551
Socio-economic factors							
Higher social inequality				*		*	
% foreign residents				*			
Mean education levels				*		*	
Mean income levels				*		*	
% blue collar workers		*		*		−0.180	
% professional workers		*		*			
Youthful population			−0.195	*			

* No data. This table only shows relationships that are 0.10 statistically significant in analysis of variance tests (as 'Prepared to break rules to help people' it was excluded from computations). The variables in the left-hand column are explained in the Appendix.

interest groups examined, except for Germany, wherein low-income groups are unusual in seeing a close relationship between activity rate and municipal responsiveness (e.g. Table 4.5, Table 4.6). Other than this, the most obvious inter-municipal distinction in terms of responsiveness to interest groups came from the strength of local political party organizations. In places where local parties are regarded as strong organizations (see Clark & Hoffmann-Martinot, 1998), responsiveness to interest groups is higher. Note that this applies not just to business groups (Table 4.3) but also to low-income groups (Table 4.5) and to neighborhood groups (Table 4.6). Note also that in a distribution of generally weak direct associations with municipal leader evaluations, this is one of the few measures for which an association of strength exists (e.g. Table 4.4). But to emphasize a critical point, the reader should recall that these evaluations are not made across nations but for municipalities in the same countries. In terms of their cross-national standing, local parties in the USA, Canada and Australia are considerably weaker than their counterparts in France, Japan, Germany and the U.K. (Clark & Hoffmann-Martinot, 1998, p. 112). As one

Table 4.5 Municipal Responsiveness to Low-Income Groups – Above/Below National Mean

Variables	Australia	Canada	Germany	Japan	UK	USA
Governance factors						
Parties more on political right			+		–	
Leaders favor more public spending	+				+	
Favor redistributive spending					+	
Socially conservative leaders			*			+
Strong local party organizations	+	+	+	+	+	+
% female councilors						
Prepared to break rules to help people		+	*			
Prepared to take unpopular decisions	+		*		–	
Importance of media	+	+	*			
General interest group activity	+	+	+	+	+	+
Socio-economic factors						
Higher social inequality	+		*		*	+
% foreign residents			*			
Mean education levels	–		*		*	–
Mean income levels	–		*		*	–
% blue collar workers	+	*	*	–		+
% professional workers	–	*	*			–
Youthful population			*		–	–

* No data. Information of responsiveness to low-income groups was not available for France. This table only shows relationships that are .10 statistically significant in analysis of variance tests (as 'Prepared to break rules to help people' is a binary variable, chi-squared was used). The variables in the left-hand column are explained in the Appendix.
+ Here places with above the national mean average responsiveness to this interest group record higher scores for the variable in the left-hand column than places below the national mean average.
– Here places with above the national mean average responsiveness to this interest group record lower scores for the variable in the left-hand column than places below the national mean average.

country is not compared with another (the same applies for all the tables in this chapter), what this set of results brings to the fore is intra-national variation. Seen most evidently for low-income groups (Table 4.5), these results show that amongst municipalities in the same nation, those with strong party organizations tend to be more responsive to interest groups than others. Lest the reader is tempted to associate party strength with political ideology here, it is worth noting that the party ideology variable is a poor discriminator of municipal responsiveness. Notably, for instance, where it does score, results are more likely to be surprising; as with parties of the right being more inclined to respond to low-income groups and less inclined to respond to business interests in Germany. Only in the U.K., with a left-wing bias in positive responses to low-income groups, do we find a predicted (ideology) pattern. As with the general

Table 4.6 Municipal Responsiveness to Neighborhood Groups – Above/Below National Mean

Variables	Australia	Canada	France	Germany	Japan	UK	USA
Governance factors							
Parties more on political right				+			
Leaders favor more public spending							
Favor redistributive spending				+		+	
Socially conservative leaders				*			+
Strong local party organizations	+			+	+	+	+
% female councilors			*	–		–	
Prepared to break rules to help people	+	+	*	*			
Prepared to take unpopular decisions				*			+
Importance of media	+	+	*	*			
General interest group activity	+	+	+		+	+	+
Socio-economic factors							
Higher social inequality	+			*		*	
% foreign residents				*			
Mean education levels				*		*	
Mean income levels		+		*		*	
% blue collar workers		*		*			
% professional workers		*		*			
Youthful population				*			

* No data. This table only shows relationships that are 0.10 statistically significant in analysis of variance tests (as 'Prepared to break rules to help people' is a binary variable, chi-squared was used). The variables in the left-hand column are explained in the Appendix.
* No data. This table only shows relationships that are 0.10 statistically significant in analysis of variance tests. The variables in the left-hand column are explained in the Appendix.
+ Here places with above the national mean average responsiveness to this interest group record higher scores for the variable in the left-hand column than places below the national mean average.
– Here places with above the national mean average responsiveness to this interest group record lower scores for the variable in the left-hand column than places below the national mean average.

trend in results obtained here, a similar picture is found when taxpayer and minority group responsiveness is examined. That party strength must be distinguished from political party ideology is made clear when we identify the uneven relationship of these two measures across nations. Thus, with a negative sign indicating that left-wing parties have stronger party organizations, the values scored when party strength is correlated with party ideology run from Australia (–0.598), through France (–0.314), Canada (–0.249), the USA (–0.243), the U.K. (–0.064), Japan (–0.051), to Germany (0.071). Excluding France, this suggests that, within countries with an overall tradition of stronger local parties (U.K., Japan and Germany), localities with stronger party organizations are as likely to be controlled by the political left as the political right (hence insignificant correlations for UK, Japan and Germany). In nations that generally have weaker political organizations, those localities that have relatively strong parties are more inclined to have a controlling party from the political left.

This point aside, there is little by way of consistent results to draw attention to. What stands out more is the absence of support for prior expectations. Thus, municipalities that revealed elements of NPC conditions are not distinguished by being more responsive to interest groups, whether these be allied to traditional lines of class division (Table 4.3, Table 4.5) or whether they represent potentially more 'neutral' class positions (Table 4.6). Note, for example, that in Australia and the USA responsiveness to low-income groups reveals noteworthy traces of class politics, with lower levels of response from municipalities whose population is comprised of those with higher income levels, more education, and a stronger professional and managerial base. We should however note that Clark provides some insight on this, when noting that weak party organization allows for a stronger transposition of socio-economic inequalities onto municipal policies (Clark & Hoffman-Martinot, 1998, p. 144, p. 133). A second point about the peculiarities of weak national party organization should be noted, which relates to the hierarchy principle of NPC. This posits that efforts to induce political change tend to be more notable when social hierarchies are stronger. As represented by the Atkinson index for educational inequality, Table 4.5 reveals that higher levels of social inequity are associated with more responsive governmental reactions to low-income groups in Australia and the USA. This is an important point, for it supports the idea that where party organization is weak this leads to a stronger transposition of socio-economic effects onto municipal policy.[7] However, in the case of Table 4.5, this relationship is only noteworthy in terms of cross-national comparisons, given that Australia and the USA (with Canada) in general have weak party systems (Figure 1.1). Within even these nations, the dominant relationship for strong party organization is that it is associated with greater responsiveness to low-income groups (Table 4.5).

Before dwelling on these results too much, we should extend the analysis by considering group responsiveness once activity rates are taken into account. For low-income groups the relationships identified are show in Table 4.7. In terms of redistributive politics, the most favorable category in this table is for municipalities in which responsiveness to low-income groups is high even though activity rates are below the national average (category R). This is not to decry situations in which both activity and responsiveness rates are above the national average, for municipal councilors can play an important part in encouraging action from interest groups. Nevertheless, if we wish to draw out the extreme positions, high responsiveness in the face of low activity levels is an appropriate category to focus on, comparing this to high activity rates that yield low levels

Table 4.7 Activity and Responsiveness to Low-Income Groups

Variables	Australia	Canada	Germany	Japan	UK	USA
Governance factors						
Parties more on political right			B		A	
Leaders favor more public spending						
Favor redistributive spending						B
Socially conservative leaders			*			R
Strong local party organizations	B	B	B	B		B
% female councilors						
Prepared to break rules to help people			*			
Prepared to take unpopular decisions			*			
Importance of media	R	B	*	B		A
Socio-economic factors						
Higher social inequality		A	*		*	B
% foreign residents	N	A	*			
Mean education levels			*		*	N
Mean income levels		A	*		*	N
% blue collar workers	R	*	*			R
% professional workers	N	*	*			N
Youthful population	N		*		N	

* No data. Information of responsiveness to low-income groups was not available for France. These results are from analysis of variance tests using a 0.10 cut-off (as 'Prepared to break rules to help people' is a binary variable, chi-squared was used). The letters represent the category recording the highest value for the variable in the left-hand column. The four categories are: N – low-income groups are less active than the national average and municipal responsiveness to them is below the national average; A – low-income group activity is above the national average but responsiveness to these groups is below it; R – the activity rate is below the national average but municipal responsiveness to low-income groups is greater than the national average; and, B – both activity and responsiveness rates for low-income groups are above the national average.

of municipal response. Viewed in this light, low-income groups are most likely to receive a favorable response in the USA in cities with a large blue collar population, in which local leaders record high levels of social conservatism. Most clearly, these features are associated with traditional blue-collar regimes. A similar relationship is recorded for Australia and Canada, with low responsiveness if cities have professional, well-educated populations (in both countries the poorest response rate relative to activity levels was for cities with a high component of college educated people). In Australia, as in the U.K., a younger population is allied to poor rates of response to low-income groups, although in the U.K. this is heightened if the political right rules the local council. In one regard, however, the U.K. is exceptional, for whereas in all other countries strong parties are associated with high rates of low-income group activity and a noteworthy municipal response to the same, in the U.K. party organization is not linked to differences in the activity-response relationship. For the U.K., party ideology is more important than local party organizational strength in municipal reactions to low-income groups.

When other interest group activity is exposed to the same analysis, a similar pattern is observed, in the sense that few of the variables which have been used to account for differences in municipal performance show themselves to be statistically significant discriminators of activity-response categories. Taking business groups by way of illustration, out of the variables listed in Table 4.7, the number recording statistically significant R or A categories only ran to six for the USA, with none for each of Australia, Germany and Japan (Canada and the U.K. recording one each and France two). Exploring some of these relationships would require a different paper than that provided here. This is seen in the importance of the local media being linked to low response rates to business in the USA, to high business responsiveness in Australia, to low activity and response levels in the U.K. and to above average activity and response rates in Canada. Similarly, strong local parties were equally likely to be associated with high and low responsiveness compared to activity levels in the USA. Yet party organization otherwise only scored in Japan and the U.K., where it was more closely linked to positive responses to business groups (albeit accompanied by high activity levels in the U.K.).

In comparing the fortunes of 'advantaged' and 'disadvantaged' groups, the analysis so far has not made direct comparisons. Clearly, if there is a real contrast between these categories we should see this when direct comparison is made between the fortunes of these types of interest group. This theme has been explored here by comparing situations in which interest groups receive above and below the national average responsiveness scores. By comparing two sets of interest groups this yields a fourfold classification for municipalities. Thus, for business and low-income groups, we have municipalities in which: (a) responsiveness to both business and low-income groups are below the national average; (b) responsiveness to business groups is above but to low-income groups is below the national average; (c) responsiveness to business groups is below, whereas for low-income groups it is above the national average; and (d) responsiveness to both sets of groups is above the national average. When these categories are subjected to investigation via analysis of variance, the results show that strong party organization and the overall level of group activity are the most important discriminators, with the importance of the local media in covering local political events also scoring highly in the three weak party states of Australia, Canada and the USA (Table 4.8). In these cases, strong party organization, media importance and level of group activity were primarily associated with both low-income and business groups being more prone to be responded to. Apart from these three basic conditions, few of the variables examined in this study revealed noteworthy relationships across nations in terms of relative responsiveness rates. For single nations the picture raises some points of interest, although governance and socio-economic conditions mostly do not distinguish between uneven municipal responsiveness. Nonetheless, it is worth noting the continuing appearance of a class politics factor in the USA. Thus, in terms of low-income groups being more prone to receive a good hearing from their city council, this is most noteworthy where there is a high blue-collar presence. On the other side of the relationship, business groups receive 'disproportionate' attention where the professional and managerial population is large, where there is a relatively young population, where mean education levels are higher, and where average incomes are greater. This picture appears in a more diluted form in Australia as well (Table 4.8). In the U.K. these relationships appear in a different form. Here relative gains for business groups owe less to underlying socio-economic conditions than to the right-wing ideology of the governing party (with councils under left-wing control revealing the reverse relationship – strong responses to low-income groups and weak ones to business groups).

Comparison between municipal responsiveness to business organizations and minority groups provides further confirmation of some of these trends. In the USA, for example, the most distinctive result is for business organizations to receive more favorable municipal responses in cities with higher mean income levels, more formal educational achievement, a greater share of the population in professional and managerial occupations, a larger share of the population aged 35–44 and where the council is controlled by the political right. These signify a traditional class politics dimension to municipal responsiveness. But where educational inequality is higher, so hierarchical relationships are more marked, where the media is considered to have an important role in local affairs, and where political parties are better organized, minority interests receive responsiveness ratings that are much higher than those of business groups. In terms of the NPC, this shows that the linkage of hierarchy with municipal responsiveness favoring minority groups is worth noting. To put this in context, it should be noted that the direct relationship between the Atkinson educational inequality measure and overall rates of municipal responsiveness to interest groups is not high. Thus, Pearson correlation coefficients range from a low positive figure for the USA (0.11) to a low negative figure for Japan (–0.12), with Australia having a weaker positive score and Canada and France recording negative coefficients close to zero (this variable was not available for Germany and the U.K.). We are not dealing with a situation in which cities with high levels of social inequality are 'naturally' more responsive to interest groups.

Table 4.8 Responsiveness to Both Business and Low-Income Groups

Variables	Australia	Canada	Germany	Japan	UK	USA
Governance factors						
Parties more on political right			L		C	
Leaders favor more public spending						
Favor redistributive spending					B	
Socially conservative leaders			*			
Strong local party organizations	B	B	B	B	C	B
% female councilors					N	N
Prepared to break rules to help people			*			
Prepared to take unpopular decisions			*			
Importance of media	B	B	*			B
General interest group activity	B	B	L	B	B	B
Socio-economic factors						
Higher social inequality	N-L-B		*		*	B
% foreign residents			*			
Mean education levels	C		*		*	C
Mean income levels			*		*	C
% blue collar workers		*	*			L
% professional workers	C	*	*		C	C
Youthful population			*			

* No data. Information of responsiveness to low-income groups was not available for France. These results are from analysis of variance tests using a 0.10 cut-off (as 'Prepared to break rules to help people' is a binary variable, chi-squared was used). The letters represent the category that recorded the highest value for the variable in the left-hand column. The four categories are: N – municipal responsiveness to business and low-income groups is below the national average; L – responsiveness to low- income groups is above the national average but below it for business groups; C – responsiveness to low-income groups is below the national average but responsiveness is greater than the national average for business groups; and, B –responsiveness rates are above the national average for both business and low-income groups.

That the U.S. pattern of more favorable responses for minority groups is confirmed for Australia, which commonly matches U.S. patterns in diluted form, offers a further pointer to a sensitivity to minority interests in such cities, especially when compared with the tendency for these places to be more responsive to business interests than to low-income groups (Table 4.8). Beyond this, once we take out the common trend for responsiveness to both business and minority interests to be related to high levels of overall interest group activity and strong party organization, there is little pattern in the relationships identified. In both Australia and the U.K. (as with the USA), political parties on the right are more inclined to favor business over minority interests (this relationship only existing for the U.K. when the contrast was between business and low-income groups; Table 4.8). Putting this point the other way around, in these countries parties on the political left appear more willing to respond to minority interests than they do to low-income group representations. This again offers some conditioning to simplistic class politics interpretation of events, albeit given the generally weak pattern of relationships that are found, no interpretive lens is convincingly supported by the analysis undertaken.

In this regard the situation for neighborhood groups turns out to be surprising.[8] When compared with business groups, too few municipal leaders rate responses to neighborhood groups to be sufficiently different from that afforded to business interests for the same analysis as presented in Table 4.8 to be undertaken. However, comparison with responsiveness to taxpayer groups was feasible. Here, responses to neighborhood organizations again revealed something of a nuanced municipal response to different interest lobbies. For example, in the USA, taxpayer groups are most likely to be relatively favored where municipal leaders are socially conservative, when overall interest group activity is high, where inequality in educational attainment is large and where party organization is strong. As neighborhood groups were examined in anticipation that they might lack a clear social class association (as groups could represent neighborhoods of varying social compositions), it is noteworthy that measures of class politics are not influential here. However, the results do not provide unambiguous support for NPC ideas. It is perhaps to be expected that socially conservative leaders will resist new social organizations, like neighborhood groups, but the association of hierarchy (educational inequality) with greater responsiveness to taxpayer groups points more to traditional 'conservative' politics (especially when reinforced by strong party organization). Outside the United States, however, these municipal characteristics are little related to differences in municipal responsiveness. The few shared relationships that do exist are that hierarchy also favors taxpayer groups in Australia and Canada, and that strong party organization is associated with taxpayer groups getting a favorable hearing in Australia (in Germany and Japan strong parties are linked to favorable responses for both taxpayer and neighborhood groups). Otherwise the most notable results are that local leaders report higher responsiveness rates for both groups in cities where interest groups are generally active and, for Australia and Canada at least, where the media has an important role in local politics.

As a final consideration, we draw on the ideas Clark & Ferguson (1983) developed 20 years ago concerning a shift in public attitudes towards a new fiscal populism. One element of this, which has been built into New Political Culture theorizing, is that leaders who adopt this policy stance downplay the role of organized interest groups. As a counterpoise, they are expected to be more responsive to general citizen interests – to cultivating the provision of public goods as opposed to catering to demands for private goods. To explore this issue, responsiveness to the citizenry (as opposed to interest groups) was examined.[9] Although data only existed for Australia, Canada, France and the USA on responsiveness to citizens, even more than neighborhood groups, responsiveness to images of citizen preferences have the potential to be less closely linked to social class divides (note how Walter finds that politicians in Stuttgart had a more accurate perception of general citizen preferences than of members of their own political party – Chapter Six). However, as Sellers (Chapter Five) warns us, the structural framework of values and legal requirements within nations does impose significant restraints on progressive politics. Intriguingly in this regard, in both France and the USA, none of the governance or socio-economic measures examined here distinguished between municipal responses that fell above or below national average scores.[10] Moreover, contradicting NPC expectations, in Canada citizen responsiveness was allied to less social inequity (hierarchy), while in Australia it was mostly distinguished by leaders preferring more public spending in general and, especially, more on redistributive services. Emphasizing how weak relationships were between city governance/socio-economic circumstances and responsiveness to citizens, of the 68 tests run for the four nations examined, just six recorded 0.10 a statistically significant discriminating effect.

CONCLUSION

Analysts regularly report that citizens are losing faith with political leaders. In country after country this is seeing expression in the dealignment of electorates, with younger populations being particularly inclined to shed (or never gain) attachments to political parties (e.g. Biorco & Mannheimer, 1995). As expressed in public opinion polls, citizens declare that their views on politics and their own citizenship are changing. In particular, citizens are prone to see citizenship as embodying horizontal rather than vertical relationships (Ashford & Timms, 1992). These changing views underpin the emergence of new political cultures, in which traditional associations and values are brought into question. In democratic nations, we might expect governmental actions to draw in changed citizen values so policy re-orientations occur. In this light, we might expect municipal responsiveness to different citizen groups to change. One suggestion that such changes have not gone far comes from examining local leaders assessments of their responsiveness to interest groups. Taking the one interest group leaders reported they responded to the most, business groups were most favored in Canada, Japan and the USA, with neighborhood associations scoring more favors in Australia, France and Germany, and low-income groups coming out on top in the UK (Table 4.1). Only in Australia (at 41.1) did the general citizenry score a higher response rate than these special interests. As neighborhood groups cannot be tied in a straightforward manner to traditional class politics, this array of responses does at least indicate potential for new orientations in municipal responsiveness. But, as Clark (1994b, 1996) warns us, if a new political culture is penetrating the mind-sets of municipal leaders, as yet this is transferring itself into public policies in a diluted form. The same can be said for responsiveness to interest groups.

Perhaps this owes something to the link between citizen satisfaction and public policy actions, for in some countries at least citizens are satisfied with the performance of local officials if they are happy with local public services (Rose & Pettersen, Chapter Two; Putnam, 1994). In a context in which citizens often seem confused about what local governments actually do (e.g. Lynn, 1992), and there is a weak relationship between 'objective' measures of service provision and citizen satisfaction with the same (Birgersson 1977; Gaziel 1982), perhaps we should not expect considerable citizen pressure on municipal governments. Indeed, as various researchers have pointed out (e.g. Piven & Cloward, 1979; Miranda & Walzer, 1994), some of the disgruntlement that led to protest activity in the 1960s and 1970s has now been incorporated into mainstream considerations during policy determination, so their presence is less overt than in the past. But while this line of argument might be thought to help explain some elements of the unpatterned nature of municipal responsiveness, this explanation cannot be taken too far (even accepting that elements of 'incorporation' cannot be measured easily). This is because the responsiveness local leaders were asked to define concerned municipal reactions to the spending preferences of interest groups, not to their overt demands for action. If incorporation of previous demands has occurred, then this should have been manifest in leader evaluations. That said, Germany apart, one of the clearest results from this study is that the overall level of interest group activity is positively related to favorable municipal responses to interest group preferences. Also important is the presence of strong party organizations, which appears to be closely allied to interest groups of all kinds receiving a supportive municipal response. Significantly, inter-municipal differences in party strength make their presence felt in countries that as a whole have weak party systems; whether in weak or strong party systems, inter-municipal variation in organizational strength is an influential force in municipal responsiveness to interest groups. Set against this, only in a few countries does party ideology make an impression on responsiveness to interests groups (most evidently in the U.K. and Germany).

Yet, irrespective of intra-national differences in party effects, there are also similarities in responsiveness patterns across states with generally weak party systems (viz. Australia and the USA). As Clark argues, a key consequence of weak party organization is that socio-economic conditions are more directly transposed onto policy choices (Clark & Hoffmann-Martinot, 1998), so socio-economic circumstances have closer associations with uneven responsiveness to interest groups than elsewhere.[11] In this regard, the combination of stronger pro-business orientations in municipalities that are richer, with more professional workers, and higher education levels, points to the continuance of lines of division based on social class. This point is emphasized when we note how U.S. cities with a large blue-collar workforce show a greater disposition to respond positively to low-income groups than to business interests (Table 4.8). Were a new political rationale dominant, arguably such underlying features should be associated with equal responsiveness across groups. Of course, the fact that few close associations were found between interest group responsiveness and both governance and socio-economic conditions might lead some to assume that 'equality' exists. This, however, fails to recognize the magnitude of unevenness in municipal responses to interest groups. Table 4.1 might suggest that there is not enormous variability in responsiveness to interest groups across most of the nations examined here. However, within these nations there are huge disparities in responsiveness to the same groups.[12] This brings out the importance of comparing responsiveness both across nations and within them. In this chapter, this point has been most obvious from the manner in which strong party organizations are allied with dissimilar impacts on interest group responsiveness at the inter-local and the cross-national level (note that high inter-municipal disparities in party strength are especially notable in nations with generally weak party systems; Figure 1.1). As Sellers (Chapter Five) points out, what is considered "progressive" (or "reactionary") in one nation can be viewed in a quite different light cross-nationally. To understand better how responsiveness is linked to governance and socio-economic conditions, detailed evaluation is required of both variation across localities and how this is tied to the peculiarities of translocal forces within each nation.

NOTES

1. An illustrative but pointed indicator of this is Rabin's (1987, p. 215) observation that acquisition of the rights of way in Osage, West Virginia, for Interstate 79, eliminated every single black occupied dwelling in the town, even though the black share of the town's population was one-third.

2. Oberschall (1973, p. 214) makes this point forcefully, first by noting that "... a negatively privileged minority is in a poor position to initiate a social protest movement through its own efforts alone," before going on to draw out the implications of this statement for 1960s civil rights action: "There can be little doubt that massive outside support and the loosening of repressive social control brought about by increasing federal government involvement and support for civil rights created the condition making possible the mobilization of the black community" (pp. 218–219).

3. The Fiscal Austerity and Urban Innovation (FAUI) database is used for this chapter. This is comprised of questionnaire returns from municipal leaders in more than 7,000 localities (see Clark & Hoffman-Martinot, 1998, pp. 168–191). Other empirical research using this database includes Clark (1994a, 1996).

4. The commonly presented argument that one of the hallmarks of Thatcherism was that interest groups were given short-shift needs to be treated with care in this regard, for while some interest groups were barely given the time of day (trade unions being an obvious example), others were more favorably treated (e.g. Kavanagh, 1987; MacInnes, 1987).

5. As one illustration, presentations in Gabriel and associates (2000) largely fail to find such data for civil associations, even in detailed studies of major European cities.

6. The empirical base for this analysis was restricted to seven nations, as these provided the most complete lists of common variables that could be used to explore interest groups and local political and socio-economic conditions.

7. Although this effect does not come across strongly for Canada, another nation with generally weak political part organizations, two of the critical socio-economic variables that bring out this point in Australia and the USA were not available in the data set for this nation (Table 4.5).

8. To qualify this point somewhat, we remind the reader that the questionnaire item in the database asked about responsiveness to spending preferences. In many cases neighbourhood group preferences are less about spending than about land-use and facility provision; albeit, as with the school closures investigated by Bondi (1988), financial considerations might be critical to a governmental decision that generates interest group activity.

9. 'Citizens' is a difficult analytical category, as leaders can have different images of what they mean by 'citizens'. Some elected officials see 'citizens' as those outside parties or well-established interest groups. But those making representations at (say) public meetings do not necessarily indicate if they are 'independent' or represent an organized group. Municipal leaders also appear to see 'citizens' more as members of groups seeking increased spending than as taxpayers, as illustrated in expectations that citizens want to 'spend more' (Clark, 1994a), when citizens often prove to be more fiscally conservative than municipal leaders (e.g. Clark & Rempel, 1997).

10. Illustrating the weakness of the relationships obtained, when raw responsiveness scores were used in Pearson correlation computations, the highest coefficient in each country for responsiveness to citizens was for leader support for more municipal spending in the USA ($r = 0.09$), the percent of the municipal population aged 25–34 in Canada (–0.22), the strength of local political parties in France (0.14) and municipal leaders favoring more expenditure on public services in Australia (0.29).

11. That said we should note that Australia and the USA are not equally placed in terms of their responsiveness to business groups, although in both countries business groups receive amongst the most favorable treatment from municipalities (Table 4.1). We should also note here that Canada has a weak political party system but does not reveal an array of relationships comparable with the Australian and U.S. cases. This requires further investigation, since two of the key measures with social class implications that were used in this study – percent blue collar workers and percent professional and managerial workers – were not available for Canada.

12. Taking three groups to illustrate this point, the coefficient of variation for responsiveness score for low-income groups ranged from 0.498 (Germany) to 1.010 (USA),

from 0.580 (Japan) to 1.506 (France) for taxpayer groups and from 0.422 (Germany) to 0.916 (UK) for business groups.

REFERENCES

Abramson, P. R., & Inglehart, R. (1995). *Value Change in Global Perspective*. Ann Arbor: University of Michigan Press.
Ashford, S., & Timms, N. (1992). *What Europe Thinks: A Study of Western European Values*. Aldershot: Dartmouth.
Atkinson, A. B. (1975). *The Economics of Inequality*. Oxford: Oxford University Press.
Biorcio, R., & Mannheimer, R. (1995). Relationship Between Citizens and Political Parties. In: H-D. Klingemann, & D. Fuchs (Eds.), *Citizens and the State*, (206–226). Oxford: Oxford University Press.
Birgersson, B. (1977). The Service Paradox: Citizen Assessment of Urban Services in 36 Swedish Communes. In: V. Ostrom & F. P. Bish (Eds.), *Comparing Urban Service Delivery System: Structure*, (243–267). Beverly Hills: Sage.
Bondi, L. (1988). Political Participation and School Closures. *Policy and Politics 16*, 41–54.
Button, J. W. (1978). *Black Violence: Political Impact of the 1960s Riots*. Princeton, New Jersey: Princeton University Press.
Caute, D. (1988). *Sixty-Eight: The Year of the Barricades*. London: Hamish Hamilton.
Chekki, D., & Toews, R. T. (1987). *Organized Interest Groups and the Urban Policy Process*. Winnipeg: University of Winnipeg Institute of Urban Studies Report 9.
Clark, T. N. (Ed.). (1994a). *Urban Innovation*. Thousand Oaks, California: Sage.
Clark, T. N. (1994b). Race and Class Versus the New Political Culture. In: T. N. Clark (Ed.), *Urban Innovation*, (21–78). Thousand Oaks, California: Sage.
Clark, T. N. (1996). Structural Realignments in American City Politics: Less Class, More Race, and a New Political Culture. *Urban Affairs Review 31*, 367–403.
Clark, T. N., & Ferguson, L.C. (1983). *City Money: Political Processes, Fiscal Strain and Retrenchment*. New York: Columbia University Press.
Clark, T. N., & Hoffmann-Martinot, V. (Eds.). (1998). *The New Political Culture*. Boulder, Colorado: Westview.
Clark, T. N., & Rempel, M. (Eds.). (1997). *Citizen Politics in Post-Industrial Societies*. Boulder: Westview.
Dearlove, J. (1974). The Control of Change and Regulation of Community Action. In: D. Jones & M. Mayo (Eds.), *Community Work One*, (22–43). London: Routledge and Kegan Paul
Doherty, B. (1998). Opposition to Road Building. *Parliamentary Affairs 51*, 370–383.
Domhoff, G. W. (1975). Social Clubs, Policy-Planning Groups and Corporations: A Network Study of Ruling Class Cohesiveness. *Insurgent Sociologist 5*(3), 173–184.
Duncan, S. S., & Goodwin, M. (1988). *The Local State and Uneven Development: Behind the Local Government Crisis*. Cambridge: Polity.
Elkin, S. L. (1987). *City and Regime in the American Republic*. Chicago: University of Chicago Press.
Fried, M. (1966). Grieving for a lost home: the psychological costs of relocation. In: J. Q. Wilson (Ed.), *Urban Renewal*, (359–379). Cambridge, Massachusetts: MIT Press.
Gabriel, O. W., Hoffmann-Martinot, V., & Savitch, H. V. (Eds.). (2000). *Urban Democracy*. Opladen: Leske and Budrich.

Gaziel, H. (1982). Urban Policy Outputs: A Proposed Framework for Assessment and Some Empirical Evidence. *Urban Education 17* 139–155.
George, J. (1999). *A Rural Uprising: The Battle to Save Hunting with Hounds.* London: JA Allen.
Griggs, S., Howarth, D., & Jacobs, B. (1998). Second Runway at Manchester. *Parliamentary Affairs 51*, 358–369.
Hahn, H. (1970). Ethos and Social Class: Referenda in Canadian Cities. *Polity, 2*, 295–315.
Hajnal, Z. L., & Clark, T. N. (1998). The Local Interest-Group System: Who Governs and Why?. *Social Science Quarterly 79*, 227–241.
Hampton, W. (1987). *Local Government and Urban Politics.* London: Longman.
Hart-Davis, D. (1997). *When the Country Went to Town: The Countryside Marches and Rally of 1997.* Ludlow: Excellent Press.
Hays, S. P. (1964). The Politics of Reform in Municipal Government in the Progressive Era. *Pacific Northwest Quarterly 55*, 157–169.
Heath, A., Jowell, R., & Curtice, J. (1985). *How Britain Votes.* Oxford: Pergamon.
Heidenheimer, A. J., Heclo, H. H., & Adams, C.T. (1990). *Comparative Public Policy.* Third Edition. New York: St Martin's Press.
Hunter, F. (1953). *Community Power Structure.* Chapel Hill: University of North Carolina Press.
Kavanagh, D. (1987). *Thatcherism and British Politics: The End of Consensus?.* Oxford: Oxford University Press.
King, R. (1979). The Middle Class in Revolt?. In: R. King, & N. Nugent (Eds.), *Respectable Rebels*, (1–22). London: Hodder and Stoughton.
Lipsky, M. (1968). Protest as a Political Resource. *American Political Science Review 62*, 1144–1158.
Lowe, P. D., & Goyder, J. (1983). *Environmental Groups in Politics.* London: Allen and Unwin.
Lowe, P. D., Clark, J., Seymour, S., & Ward, N. (1997). *Moralizing the Environment: Countryside Change, Farming and Pollution.* London: UCL Press.
Lynn, P. (1992). *Public Perception of Local Government: Its Finance and Services.* London: HMSO.
MacInnes, J. (1987). *Thatcherism at Work: Industrial Relations and Economic Change.* Milton Keynes: Open University Press.
Miranda, R. A., & Walzer, N. (1994). Growth and Decline of City Government. In: T. N. Clark (Ed.), *Urban Innovation,* (146–166). London: Sage.
Newby, H.E. (1980). Urbanization and the Rural Class Structure. In: F. H. Buttel, & H. E. Newby (Eds.), *The Rural Sociology of the Advanced Societies,* (255–279). London: Croom Helm.
Oberschall, A. (1973). *Social Conflict and Social Movements.* Englewood Cliffs, New Jersey: Prentice-Hall.
Page, E. C., & Goldsmith, M. (1987). Centre and Locality: Explaining Crossnational Variation. In: E. C. Page & M. Goldsmith (Eds.), *Central and Local Government Relations,* (156–168). London: Sage.
Piven, F. F., & Cloward, R. A. (1979). *Poor People's Movements: How They Succeed and Why They Fail.* New York: Vintage.
Putnam, R. D. (1994). *Making Democracy Work: Civic Traditions in Modern Italy.* Princeton, New Jersey: Princeton University Press.
Rabin, Y. (1987). The Roots of Segregation in the Eighties: The Role of Local Government Actions. In: G. A. Tobin (Ed.), *Divided Neighborhoods: Changing Patterns of Racial Segregation,* (208–226). Newbury Park, California: Sage Urban Affairs Annual Review 32.
Sjoberg, G., Brymer, R. A., & Farris, B. (1966). Bureaucracy and the Lower Class. *Sociology and Social Research 50*, 325–337.
Stokman, F. N., Ziegler, R., & Scott, J. (Eds.). (1985). *Networks of Corporate Power.* Cambridge: Polity.

Stone, C. N. (1980). Systematic Power in Community Decision-Making. *American Political Science Review 74*, 978–990.

Swanstrom, T. (1985). *The Crisis of Growth Politics: Cleveland, Kucinich and the Challenge of Urban Populism.* Philadelphia: Temple University Press.

Urry, J. (1986). Class, Space and Disorganized Capitalism. In: K. Hoggart, & E. Kofman (Eds.), *Politics, Geography and Social Stratification,* (16–32). London: Croom Helm.

Verba, S. (1987). *Elites and the Idea of Equality: A Comparison of Japan, Sweden and the United States.* Cambridge, Massachusetts: Harvard University Press.

Verba, S., & Nie, N. H. (1972). *Participation in America.* New York: Harper and Row.

Weinstein, J. (1968). *The Corporate Ideal in the Liberal State 1900–18.* Boston: Beacon Press.

APPENDIX

The seven main components of the NPC approach are:

1. That the classic left-right dimension of political conflict is being transformed.
2. That social and economic issues are being explicitly distinguished, by citizens, parties and leaders.
3. That social issues have risen in salience relative to fiscal/economic issues.
4. That market and social individualism are growing as skepticism towards traditional left-right politics grows.
5. That people are questioning the welfare state more.
6. That a rise of issue politics and broader citizen participation is occurring, with the result that hierarchical political organization is in decline (new social movements as opposed to traditional party representation).
7. That these NPC views are more prevalent amongst the young, the more educated and more affluent people and localities.

These components lead to three general principles that guide a large number of theoretical propositions. The first of these principles centers on issues of *hierarchy*. Economic, political and social hierarchies are held to generate reactions against them (the leveling principle), with the expectation that as hicrarchy declines people invest energies less in anti-hierarchy movements than in a range of new forms of activity. Hence, as hierarchy declines, NPC is more evident. Secondly, there is the empowerment *principle*. Put simply, it is held that as citizens acquire more personal resources in the form of social capital (education, communication skills, experience of civic organizations, etc.), the chances of them joining issue-related lobbies or organizations increase. Their reliance on political leaders falls. Hence, the greater the skill individuals acquire the more likely they will become politically active. Thirdly, *structural* conditions are held to have an important part to play in political activity. In the nineteenth century industrialization was critical in this regard, as it transformed entire societies, while today globalization is critical, as it intensifies competition between cities, helps the spread of specific causes, helps unite non-proximate interest groups, brings actions in single localities often under an international microscope, etc. (on these three principles and associated propositions, see Clark & Hoffmann-Martinot, 1988, pp. 36–65). In terms of municipal leader values, municipal governance structures and local socio-economic conditions, these principles and their propositions lead to the deployment of the following variables to assess the prevalence of NPC and class politics (see also Clark, 1996):

Governance factors
Parties more on political right (IPARTY) The higher the score here the more the local party stands on the political right. The values range from 0 to 100. This measure was designed to permit cross-national comparisons. Explanation of the procedure involved here can be found in Clark and Hoffmann-Martinot (1998, p. 179).

Leaders favor more public spending (PRFAVG) Measures municipal leader preferences to spending more on 13 policy arenas, so distinguishes high-spending from low-spending priorities.

Favor redistributive spending (REDISTX) Similar to PRFAVG, but is couched in terms of policy fields that support the disadvantaged. In term of political ideology this reflects a classic left-right divide.

Socially conservative leaders (SOCCONS) Social issues are evaluated in terms of leader values towards abortion and sex education, which have been summed into this index. High scores on this measure represent socially conservative municipal leaders.

Strong local party organizations (SPOIX) reflects how important political parties are in listing electoral candidates, in election campaigns and in policy-making.

% female councilors (PCWOMC2) Places which are more liberal in their political dispositions are expected to be associated with higher representations of women on the local council.

Prepared to break rules to help people (IV130) This item taps leader views on the salience of respecting the rules, as opposed to achieving outcomes for citizens.

Prepared to take unpopular decisions (IV142) This is a populism measure. The second measure to tap governance processes (with IV130).

Importance of the media (MEDIA) This taps the importance of the media in elections and news coverage.

General activity of interest groups (GRPACT) Municipal leader evaluations of how active local organized groups are in their localities.

Socio-economic factors
Higher social inequality (IATKEDU5, IATKINC5) This measure was not available for Germany and the U.K., and for Canada the inequality index based on educational levels did not yield statistically significant results, whereas the income based measure did. This variable is entered to reflect the degree of inequality in local social systems, as represented by Atkinson's (1975) inequality index, as applied to educational levels and income levels. In the results presented here, education levels have been used, as income data were also not available for France and Japan.

% foreign residents (FORNSTK) This is the percentage of immigrants or foreign-born persons, depending on the character of each nation's statistics. Examining municipal responsiveness to minority groups reinforces the pertinence of this measure.

Mean education levels (IMEANEDU) This variable was not available for Germany and the U.K. It is measured at the municipal level.

Mean income levels (INCOME1) This variable also was not available for Germany and the U.K. It is measured at the municipal level.

% blue collar workers (IPCTBLUE) A classic measure of traditional class politics and political cleavage. The variable was available for all countries save Canada and Germany.

% professional workers (IOCPROF) Potentially this taps a postindustrial dimension of the municipal labor force. However, interpretation of its coefficients needs to be undertaken in conjunction with other significant relationships, as this measure can also highlight some dimensions of classic class politics (e.g. if counterpoised against relationships for blue collar workers that are significant, and in the opposite direction). Most evidently if this measure is associated with

support for business over minority and low-income groups, with blue collar representation having the opposite effect (as happened here for the USA), this appears more as a class politics dimension. This measure was not available for Canada, for which the percentage of the workforce with at least four years of college education was used.

Youthful population (IPOP3544, IPOP2534) The inclusion of this variable comes directly from the seventh NPC proposition listed above. Owing to data availability, for Australia the percentage of the population aged 25–34 was used, elsewhere the age group 35–44 was used (as in other studies using this database; Clark & Hoffman-Martinot, 1998). No age data were available for Germany.

5. TRANSLOCAL ORDERS AND URBAN ENVIRONMENTALISM: LESSONS FROM A GERMAN AND A UNITED STATES CITY

Jefferey M. Sellers

Throughout much of the advanced industrial world, measures to improve the local environment have emerged as an inextricable element of urban politics (Clark & Goetz, 1994; Sellers, 1998). A growing literature identifies these local policies with 'Progressive' or Left-oriented coalitions in the USA (Clavel, 1986; DeLeon, 1992) and Europe (Clavel & Kleniewski, 1990; Sellers forthcoming). At the same time, much of the literature on U.S. policy toward land use continues to stress broad contrasts with urban environmental policy in other advanced economies (Jackson, 1985; Nivola, 1999). To account for national differences, this work points to an array of organizational, normative, financial and other structures embedded in nation-states. Even Progressive coalitions that pursue aims contrary to those ordained from above must work out local strategies and assemble local interests that reflect the influence of these wider orders. As Progressive coalitions have brought together broadly parallel political forces around analogous local aims of environmentalism, prosperity and social justice, in both the United States and Europe, these orders reinforce a persistent, perhaps even widening, contrast between the USA and much of Europe in the politics of urban environmental policy.

In this chapter, I show how urban environmental governance under Progressive coalitions reflects this wider contrast in political and market opportunities. Since local environmental quality possesses the indivisible, equally available character of a *public good*, environmental policy might seem to offer an ideal arena for local coalitions to counter the established propensity of urban policy in the United States to rely on private solutions to public problems (Barnekov et al., 1989; Garber & Imbroscio, 1996). If successful in challenging this national tendency, Progressive coalitions in the USA could construct localized regimes of collective environmental governance similar to those in European countries.[1] My analysis will demonstrate how *translocal orders* of institutions, organizations and policy-making at higher levels of government, by defining nation-specific contexts for local policy, affect this possibility. These contexts confront urban coalitions of the Left in the USA with conditions that require environmental quality to be pursued more through *private* or divisible, *selectively* provided means, than in parts of Europe. As a result, environmentalist urban coalitions in the USA have attained not only less environmental and social justice, but worse environmental conditions overall than in countries where the translocal order fosters more public provision of environmental goods.

The experience of environmentalist coalitions in a similar U.S. and German city will demonstrate this point. My account begins with an overview of the ways that urban coalitions might link provision of environmental amenities to other ends. The next sections employ the specific comparison to illustrate the extent of cross-national differences in the outputs of regime strategies. Finally, I show how different relations between the institutionalized arenas of urban politics account for much of the contrast.

ALTERNATIVE STRATEGIES OF URBAN ENVIRONMENTALIST COALITIONS

With the growing institutionalization of environmental policy at all levels of state-society relations, ecological objectives have increasingly emerged as a focal point of coalition-building in urban arenas. Analysis of Progressive coalitions has generally associated environmental policy with distributive equity and opposition to growth (e.g. Clavel, 1986; DeLeon, 1992). However, possible combinations of environmentalism with the other policy objective, pursued by distinct types of environmentalist coalitions suggest a more complex and contingent linkage among Progressivism, equity and resistance to growth.

The environmental policies that play an increasing role in urban coalitions serve global environmental ends at the same time as they contribute to the sustainability of regional and local ecosystems. For local residents and visitors, environmental policies provide more immediate amenities that, along with related cultural policies, can enhance the quality of urban life. Although environmental measures of this sort extend from recycling to pollution control, the most consistently relevant policy for urban coalitions has centered on matters of land use. Throughout most Western cities, local policies of this sort encompass protective measures for historical and environmental preservation, maintenance of urban spatial contours, limits on auto traffic, and amenities like forests and other open spaces (Clark & Goetz, 1994; Sellers, 1998). Formal land use rules constitute only the most obvious instance of such environmental measures. Nondecisions, refusals to provide infrastructural services and indirect measures, like development impact fees, often serve analogous purposes (Altshuler & Gomez-Ibanez, 1993). Policies to establish and maintain such amenities as public parks, bicycle paths, pedestrian zones and public transit, also require initiatives that change patterns of land use.

Increasingly, these measures have joined developmental and distributive aims as the objects of urban coalitions. To one degree or another, it is usual for environmentalist aims to be reconciled with the developmental aims that predominate in many urban settings. In Europe as well as the United States, the literature on urban coalitions emphasizes efforts to attract new economic development, to encourage better performance amongst local firms or to mobilize elites around new projects (Judd & Parkinson, 1990; Le Galès, 1993; Harding et al., 1994). Even accounts that question how far growth coalitions dominate local politics in Europe reaffirm this common element (Harding, 1994; Le Galès, 1995; Strom, 1996; John & Cole, 1997). At the same time, analysts of Progressive coalitions note the importance of equity and distribution concerns to mobilization around environmental aims (e.g. DeLeon, 1992). Explicitly distributive policies extend from public provision of housing and services for lower-income people, to protection of jobs and convenient living conditions for the same residents, to redirection of profits from land development through the state (Fainstein, 1993). U.S. work on environmental justice (Bullard, 1993; Gottlieb, 1993), and European analyses of social ecology (Liepietz, 1995; Jahn, 1996), demonstrate the relevance of redistributive concerns in efforts to prevent environmental bads and provide environmental goods.

The role of environmental measures can only be understood fully in the light of the ways that Progressive coalitions, as well as growth coalitions, combine ecological objectives with other ends. New parks, bicycle paths and wooded areas need not conflict with development, as they could be incorporated into the aims of a development. Rather than simply pose obstacles to growth, preservation of nature and prevention of pollution could make a place more attractive to the workers and firms growth coalitions seek to attract. Similarly, environmental policies might serve elite groups disproportionately, as advocates of environmental justice in the United States have held. But urban environmental policies that are attentive to equity could provide essential public amenities for all, or rectify class and racial injustices in the distribution of environmental goods and bads. Even coalitions built primarily around environmental issues can gain support through attention to aims like growth and equity. A consideration of the possible ways coalitions might synthesize environmental with other aims yields three logical possibilities.

One possible combination, which I call a *Green* strategy, would emphasize environmental policies and equitable distribution but de-emphasize growth. At the same time as a Green coalition would extend levels of urban environmental amenity, it would limit special benefits for privileged groups and redistribute environmental and other benefits in favor of disadvantaged groups. Although Red-Green alliances in German cities like Frankfurt (Prigge & Leiser, 1992), alongside parallel coalitions in some U.S. cities (e.g. DeLeon, 1992), have arguably pursued this synthesis, few successful governing coalitions of this sort appear in the literature. U.S. regime theorists cast particular doubt on whether a coalition could sustain this synthesis. On the one hand, they argue, even Progressive coalitions with predominant electoral support must rely on the co-operation of local business elites with an interest in growth politics to effectuate environmental or equitable strategies (Stone, 1989; Fainstein, 1993). On the other hand, the more privileged and middle-class constituencies who provide much of the support for these coalitions maintain more interest in environmental measures than in policies addressing equitable concerns about the poor and minorities (Stone, 1993).

Parallel difficulties should confront an *Integrated* strategy. This is a strategy that promotes growth, provides environmental amenities, and addresses equitable concerns about both prosperity and environmental quality. In Europe, policy-makers and analysts who have pursued 'ecological modernization' at multiple levels of government have espoused a synthesis of this sort (e.g. Weale, 1993). In German cities, the support of the SPD for local growth within Red-Green coalitions appears to have fostered local strategies closer to this synthesis than to the Green one (Berger, 1995, pp. 279–338; Zeuner & Wischerman, 1995, pp. 201–203). Similarly, U.S. students of urban political economy portray pro-growth politics as a necessary accompaniment to other aims of urban politics (DeLeon, 1992; Fainstein, 1993; Lewis, 1996). In advanced economies, efforts to bring new development increasingly seek to attract managerial and professional workers in higher level services or technically advanced industries. Because quality of life plays an important role in the decisions of these workers about where to locate, enhancements in local environmental quality can complement other efforts to attract these workers. Like convenient shopping or good schools, public parks, pedestrian walkways, bicycle paths, a landscaped countryside and recreational facilities, all contribute to the quality of life. A well-preserved natural and urban environment enhances this value. Regulatory controls prevent new development from overburdening the physical infrastructure of roads, sewers and schools.

Yet environmental amenities and other efforts to attract privileged workers enter easily into tension with equity aims. Although new development may provide essential revenue, jobs and other resources for poorer and minority communities (Fainstein, 1993), the post-industrial employment that arrives usually aggravates disparities in income and lifestyle. Environmental amenities designed to attract the privileged can more easily do so through benefits directed at this group. Zoning and preservation can protect wealthier neighborhoods, parks can secure them public amenities, while vehicle and commercial restrictions can reduce pollution there. Selective provision can also make environmental amenities depend on purchasing power, like the environmental amenities large-lot zoning confers on residential homebuyers.

As a result, U.S. and European analysts have often portrayed *Upscale* strategies as the principal approach of most Progressive coalitions. These strategies, which Stone (1993) identifies with a 'middle class' orientation and social basis, aim to improve environmental amenities and expand development. At the same time, unlike the objectives of what Stone calls an "opportunity expansion regime," Upscale strategies de-emphasize egalitarian concerns about amenities or benefits from development for disadvantaged groups. According to the logic of Upscale strategies, the chief aim lies in bringing post-industrial growth. Policies tailored narrowly to the interests of mobile privileged workers attract new firms that assure this growth. Public environmental goods from zoning or neighborhood amenities can permit a Progressive coalition to appeal across class and ethnic lines with this strategy. But Upscale strategies ultimately demand that the disadvantaged agree to selective and private benefits for the more privileged for the sake of a minimal share in local prosperity. Although U.S. suburbs represent perhaps the clearest examples of this synthesis (Silver & Melkonian, 1995), analysts of large U.S. and German cities where Progressive coalitions have formed often point to the dominance of Upscale tendencies (DeLeon, 1992; Ronneberger, 1994). These studies suggest policies how easy it is for ecological modernization to 'go upscale'.

As regulation theorists have suggested, pressures toward either Upscale policies or marginalization of environmental concerns exist throughout advanced industrial society (Lauria, 1997). Partly because side-by-side comparative case studies of urban environmental policy remain rare, few empirical studies exist to document how much these pressures vary among countries or world regions.

Still fewer studies have analyzed the significant differences that translocal contexts of policymaking and politics make for urban coalitions and policies.[2] Such an analysis remains crucial to illuminate the opportunities that Left-leaning coalitions around environmental issues face.

CLASSIFYING THE OUTPUTS OF ENVIRONMENTALIST COALITIONS: A TALE OF TWO CITIES

Side-by-side comparison of U.S. and German environmental coalitions in otherwise similar cities can exemplify the difference national contexts make. This is illustrated here by considering the accomplishments of Left-leaning coalitions that have consolidated a dominant electoral position since the 1970s in Madison, Wisconsin, and the south German city of Freiburg im Breisgau. Faced with largely similar urban environments, economic positions, social constituencies and political traditions, and occupying similar places along the political spectrum of their respective countries, these coalitions might have been expected to yield similar results. Yet, while the coalition in Freiburg has carried out an environmentalist strategy that comes close to the Integrated type, the coalition in Madison remained Upscale by comparison. Despite the fact that, for a U.S. city, Madison is environmentalist and attentive to equity issues, it had done far less to constrain place-based disparities in environmental and other conditions, and had fallen short in environmental policy itself.

The resemblance between Madison and Freiburg makes the contrast of national contexts particularly instructive. It would be difficult to find another U.S. city where economic functions, geographic situation, social conditions and political traditions more closely approached those of a mid-sized German educational and administrative center. In both cities, concentrations of services foster business and institutional interests, as well as electoral constituencies, that are supportive of strong environmental policies. A large public university, a hospital center and regional-state administrative centers dominate both economies. Each city has also capitalized on the 'place luck' of a uniquely attractive environment to secure moderate growth: in Freiburg, the rolling wooded hills of the Black Forest; and, in Madison, the wooded lakefronts surrounding downtown. The racial divisions that hampered Progressive coalitions in many U.S. cities (e.g. Clavel, 1986) have played a marginal role in Madison. Here there is a small minority population (10% overall, with 4% African-American), which resembles the 8% foreign population in Freiburg. As in Germany, urban growth in Madison has remained more within the central city than is typical for U.S. metropolitan areas. Planning traditions more like those of Germany have helped bring this about. Nineteenth century founders planned the downtown neighborhoods of the Isthmus. Early twentieth century reformers brought elements of German planning to the city (Mollenhoff 1982). More than in most U.S. cities, the downtown continues to provide an attractive central setting for residence and shopping, analogous to the medieval old town in Freiburg.

The coalitions that had consolidated a dominant position in electoral politics in both cities by the 1990s united parallel constituencies. Since the 1970s in Madison and from the 1980s in Freiburg, political entrepreneurs, who were linked most closely to the Left, incorporated challenging groups from social movements, urban neighborhoods and the university community, alongside middle class professional constituencies, into environmentalist coalitions. In each case shifts toward environmentalist orientations in local political culture undergirded the position of these coalitions (Rosdil, 1991; McGovern, 1997).

In Freiburg, Rolf Böhme, the mayor since 1982, carried on a tradition of Social Democratic (SDP) control that extended back to 1962. Relying on strong official powers over the council and its administration, while alternating alliances with the CDU and the Greens, the mayor played the most important role in official policy. Local environmental movements emerged in the 1970s among downtown professionals and students. The movements challenged aggressive local development and gentrification, and grew with repeated nation-wide mobilizations against a nuclear reactor planned for the nearby community of Wyhl. Beginning in the 1980s, and drawing on these movements, the newly organized Green Party emerged to spearhead environmental policy on the local council. By 1989 the Greens, adding to the working-class constituencies for the SPD, enabled Left parties to secure a majority of council seats. Shortly thereafter, the Greens surpassed the Social Democrats as the largest council fraction on the Left.

In the nonpartisan political arena of Madison, Paul Soglin had assembled an electoral coalition that brought together the university community and downtown residential neighborhoods with suburban professionals from outlying neighborhoods. A prominent student activist in the antiwar protests of the 1960s, Soglin served two terms as mayor in the mid-1970s, then three further terms starting in the late 1980s. Moderately liberal mayors from 1979 to 1989 drew on much the same constituencies. Much like their counterparts in Freiburg, Soglin and the other mayors sought to accommodate planning and environmental policies with demands of the local business community for continued development. As in the German city, a council minority based mostly in downtown residential neighborhoods, the university community and active social movements, pushed persistently for environmental amenities and equitable concerns. Informally linked to the Democratic Farmer-Labor Party, this group played a role parallel to that of the Greens in Freiburg.

Each of these coalitions had consolidated local syntheses of environmental policy that stood out in their respective countries. The German press sometimes referred to Freiburg as the 'ecological capital' of the country (Obertreiss, 1988, p. 64). In part because of the environmental amenities touted in local marketing, Madison garnered multiple awards from *Money Magazine* and other sources over the mid-1990s as one of the best cities in the USA to live and work. Although the universities, and the local political economies linked to them, helped account for this leadership, close comparison of environmental indicators shows that the policies in these towns had gone beyond even those of similar university towns (Sellers, forthcoming). Freiburg had controlled new development, minimized loss of forest and agricultural land, and effectively promoted transit and other energy-efficient practices. A proliferation of plans, regulatory measures and policies in Madison from the 1970s onward proclaimed similar aims. In both settings the strength of local civic groups and widespread cultural orientations in favor of environmental values help to account for this leadership. This comparative success in each case only highlights the significance of the clear contrasts between the environmental policies of the two settings.

Despite similarities in relation to other cities in each country, policies directed at environmentalist ends in Freiburg went consistently further than in Madison. Unusually systematic records on land use for a U.S. city provide a statistical overview of this tendency (Table 5.1). As indicators of the outputs from local policy, these figures generally reflect not only the measures taken by coalitions but the effective significance of the measures implemented. Freiburg actually increased the share of its territory classified as covered by forest by nearly 8% between 1978 and 1988. This increase exceeded a small loss of agricultural land so the proportion of open space in these two categories rose by 5.2% over these 10 years. The city government thus succeeded in its goal to create more forest than new development destroyed (Freiburg, 1980, p. 54). In Madison, however, both the overall amount and the proportion of land shrank steadily over the 1970s and 1980s. Only in the provision of parkland did the Madison regime clearly exceed its German counterpart. But this result only came about because Freiburg did not count its massive city-owned forests as parks, while Madison did. Even with smaller amounts of parkland, the Freiburg city council spent $30 more per resident on parks.

The German city had also succeeded in promoting alternatives to the automobile far more systematically. Alongside influences from wider transportation policies, the urban regime had employed a range of relevant policies to shape

Table 5.1 Indicators of Strategies Toward Environmental Amenities

	Freiburg	Madison
Protection of land		
(Forested land)	65.8 km² (43% in 1980), 93.6 km² (50.8% in 1988)	
(Agricultural land)	38.25 km² (25% in 1980), 41.24 km² (22.4% in 1988)	
(All agricultural and undeveloped land)		98.8 km² (54%) (1970), 82.7 km² (41%) (1990)
Parks		
Land	4.88 km² (1988) plus 50.32 km² city forest (1980), or 31.7 cm2/resident	15.05 km² total (1990), or 7.9 cm²/resident
Expenditure	$100/resident (1995)	$70/resident (1992)
Pedestrian zones	7,500 meters (1995)	ca. 1,000 meters (1995)
Mass transit		
Types	Trolley, bus	Bus
Usage	24% (commuters to city in 1987), 26% (trips by university students in 1988)	7.6% (usage for trips to work by employed city residents in 1990)
Bicycles		
Paths	km. In 1981, 125.7 km. In 1987 (all off-street)	32.2 km. (off-street paths in 1992)
Usage	27% (overall usage for trips by residents in the late 1980s, 43% for trips by university students in 1988)	11% (for overall usage for trips by residents in the summertime 1995)
Planning administration	(1995)	(1992)
Expenditure	$33.2 million	$16.6 million
Planning staff	36	23
Enforcement staff	61	48
Land administration staff	93	15

Sources: Forested, agricultural and park land, as well as bicycle paths, for Freiburg in 1988, are from Amt für Statistik und Einwohnerwesen (1988). Forested and agricultural land, along with city forest figures for 1980 are in Stadtplanungsamt Freiburg (1980). Undeveloped land in Madison is from unpublished statistics of Dane County Planning Commission. Pedestrian zone length has been estimated from maps in Michelin Tire Company (1995, p. 303) and the city map of Madison. Expenditure and staffing figures are from city budgets for 1992 (Madison) and 1995 (Freiburg). Bicycle and mass transit usage for Freiburg are from Tressel (1990a, p. 15) and, for students, from Tressel (1990b, p. 56). Mass transit usage for Madison is from United States Bureau of the Census (1993). Bicycle paths for Madison are from Madison (1992, p. 142); usage figures for Madison are from Madison (1995, p. 47).

settlement patterns. In Freiburg, the pedestrian zone had grown by the 1990s to encompass nearly the entire downtown of the old city, or some 7,500 meters of streets. By comparison, the single passage of the State Street transit mall between the University campus and the state capitol in Madison amounted to only a modest commercial success. Efficient trolley and bus services in Freiburg provided public transit for over a fourth of commuters. In Madison, less than 8% of commuters reported using public transportation for work. Off-street bike paths in the German city grew rapidly to four times the length in Madison. Again, the German regime established routes and other facilities that fostered broader usage. Twenty-seven percent of all commuters and fully 43% of students, compared to only 11% of summertime commuters in Madison, rode bicycles.

Greater expenditure and more staff for planning administration in Freiburg reflected the more extensive efforts that produced these results. Besides those charged with enforcement, staff for permits and land use matters exceeded the size of their counterparts in Madison by a third or more. Planners themselves, but above all administrators responsible for record-keeping and acquisition of land, numbered significantly more in the German city.

Comparison with a handful of additional German university towns showed Madison to be somewhat below the average by German standards (Sellers, forthcoming). Taking into account the more ample environmental amenities provided through private means rather than directly through public policy in Madison, only partly makes up for this deficiency. On the East Side and especially on the West Side of Madison, large, privately owned lawns and woods maintained pastoral surroundings far more extensive than the smaller residential plots of well-to-do Freiburg neighborhoods. Country clubs, golf courses and exclusive residential communities, as well as city parks, interlaced these areas of private residence. The East Towne and West Towne malls, in recent years more important as regional centers of retail sales than downtown itself, offered alternative, artificial settings for congregating and shopping to the pedestrian zone of the downtown (Figure 5.1). Despite these qualifications, Freiburg had both a lower overall density of settlement (census figures show 1,167.4 persons per square kilometer in 1987 compared to 1,278.5 persons per square kilometer in Madison in 1990) and larger proportions of its land under forest than Madison.

The distribution of opportunities to take advantage of environmental amenities diverged even more than aggregate indicators. Along with schools, shopping, and transportation, differences in environmental conditions in Madison followed a clear hierarchy of access among residential neighborhoods. Persistent place-based disparities reinforced an Upscale distribution of environmental goods. Systematic separation of neighborhoods by economic means made these disparities possible.

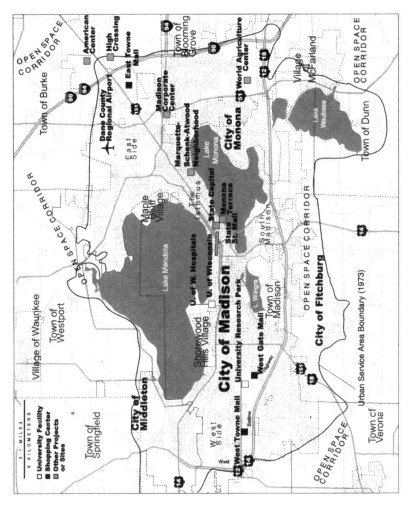

Figure 5.1: The Madison Urban Service Area.

African-Americans and the larger group of residents who live below the poverty line face levels of segregation that approach those in larger, more ethnically divided U.S. cities. Indices of residential dissimilarity stood at 31 for unemployed 'economically active' residents in relation to employed residents, at 50 for those below the poverty line as a proportion of all those for whom poverty status was determined, and at 49 for African-Americans in relation to all others (based on 1990 U.S. Census bureau STF files for block groups). These figures neared the level of 60 which Massey & Denton (1993) take as a threshold for extreme segregation. Beyond a small number of highly exclusive wealthy neighborhoods, most managerial and professional families generally lived in neighborhoods where their own relatively privileged middle class occupations predominated. In 1990, 56% of managerial and professional workers lived in census tracts where that group made up at least 40% of resident workers (United States Bureau of the Census 1993).

In the well-to-do and elite neighborhoods on the western and eastern peripheries of the city, private means alone assured the highest levels of open spaces, woods and other amenities. Large lawns, extensive woods, office parks, country clubs and private residential associations extended across these areas. The city itself contributed the major part of public parkland, nature reserves and public golf courses. The best schools, the two major shopping malls, convenient highways and a growing proportion of offices, have shifted much of the city's life outside the downtown. The master plan of the city proclaimed a need for parks and open space that is accessible to ". . . all segments of the City's population" (Madison, 1985, p. 36). It was the case that the poor in the neighborhoods of South Madison lived close to numerous parks. Yet these residents lacked the means of others to secure additional amenities for themselves. The transportation arteries, strip malls and commercial facilities that divided up South Madison restricted access even to nearby public amenities. Residents expressed fears about safety in public areas. These difficulties compounded disadvantages like the lowest ranked local schools, restricted shopping opportunities and limited access to jobs (South Madison Steering Committee, 1990).

In Freiburg, the actors of the Progressive coalition helped ease pre-existing social and economic disparities among residential neighborhoods. Doing so helped assure more equal benefits from environmental as well as other amenities. Segregation indexes for the city showed levels of less than half those for either privileged or disadvantaged groups in Madison. For unemployed 'economically active' residents in relation to employed residents, local census statistics from 1987 showed an index of dissimilarity of only 13. For foreigners in relation to the majority population, the index was only 19. (Note that in both instances the larger population of the census districts made these results more comparable

Translocal Orders and Urban Environmentalism 129

to the larger census tract figures of the U.S. Census rather than to smaller block groups). Since the 1960s, new parks, play areas, shopping facilities and transportation lines in working class neighborhoods of mass housing on the western side of the city greatly alleviated disparities between that area and the more well-to do east side. A small portion of the most privileged residents moved to outlying neighborhoods and adjacent towns along the Black Forest (Figure 5.2). The cost of housing in gentrified areas helped bring about new clusters of foreigners and the unemployed in small pockets of older, cheaper housing in industrial districts with few amenities. But the generally integrated pattern of residence for both groups, as well as the highly developed system of public

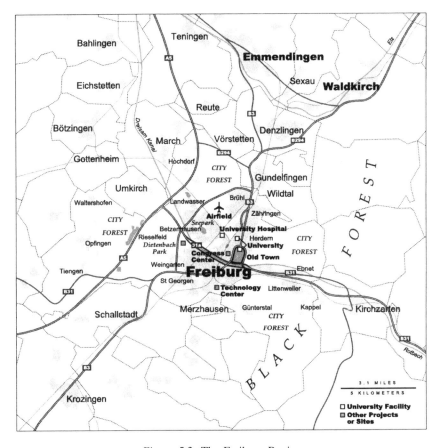

Figure 5.2: The Freiburg Region

transit, bicycle paths and walkways, assured more equal access to the benefits of local environmental policies than in Madison. Although those with college degrees made up 17% of the population aged over 25, no census district contained more than 30% of this group. Similarly, although foreigners made up 8% of the population, no district contained more than 18% (these figures are from 1987 Census, furnished by the Freiburg Statistical Office).

The policies canvassed in this brief overview highlight shortcomings that are likely to be even more apparent for the environmentalist coalitions of other U.S. cities. Where more urban and ex-urban development has taken place in outlying jurisdictions, where racial divisions have been more pervasive, and where environmentalism has lacked the local traditions of Madison, urban environmentalism faces even greater challenges. Even in Madison, the coalition must be assessed as not only Upscale by comparison with its German counterpart, but more modest in its overall environmental policies. By international comparative standards, the coalition in Freiburg belonged more to the Integrated type.

TRANSLOCAL ORDERS AND URBAN ENVIRONMENTALISM

The contrasts between urban environmentalism in Madison and Freiburg partly result from the different strategies of local officials, citizens, consumers, and businesses, as well as from the cultural orientations of local actors. But in different countries even otherwise similar urban environmentalist coalitions face distinct translocal infrastructures of governments, policies and organized representation. Typologies of 'national systems' of local government have highlighted important components of these infrastructures (Page & Goldsmith, 1987; Keating, 1991; Page, 1991; Wolman & Goldsmith, 1992). What matters for a multilevel analysis is how institutional systems and policies anchored at higher levels of the state affect the capacities and orientations of urban environmentalist coalitions. In Freiburg, the translocal order assured the Progressive coalition the means to pursue environmental ends more as collective goods. With greater capacities to limit residential segregation, the coalition also forestalled the emergence of particularistic environmental interests among more well-to-do residents and businesses. In Madison the corresponding infrastructures of local governance left the outputs of environmental policy more dependent on private or divisible means of provision. The segregation that resulted encouraged more privileged groups to develop interests in policies that favored their own neighborhoods, and limited overall provision.

Policies that preserve forests and open land, that encourage environmentally friendly means of transportation, and that conserve other natural resources, provide unmistakable public goods for ecosystems and society. At the same time, many such amenities amount to 'luxury goods' for privileged groups (Martinez-Alliez, 1995; Sellers, 1999). Privileged residents themselves may be willing to pay a premium to enjoy environmental amenities. But if these groups can provide amenities for themselves without having to do so for the remainder of society, they have less reason to work or vote for more widely distributed environmental goods. Economists refer to this particularistic or self-regarding exploitation of public policy as 'rent-seeking'. Although middle class and upper middle class constituencies have sometimes been ascribed a more neutral, less self-interested 'ethos' than ethnic working class ones (Banfield & Wilson, 1963), the bias of middle class reform toward middle class interests in many U.S. cities belies unmistakable self-regarding propensities (Bridges, 1998). Where middle and upper middle class residents make up a sizeable portion of the community, both development markets and political coalitions are likely to reflect and reinforce the preferences of this group. Not only businesses concerned directly with development, but those interested in the wider economic prosperity of a city, can profit from a vision of development that efficiently satisfies the preferences of these well-to-do residents. Perceptions of translocal markets reinforce the bias of policy in favor off privileged groups. Whatever the national context, urban environmentalist coalitions typically rely on these same middle- and upper-middle-class residents for electoral support. At the same time, urban policymakers must implement environmentalist policies in cooperation with business in need of middle and upper middle class clienteles. For both reasons, progressive coalitions often find it useful to secure more selective or privatized environmental goods over more collective ones.

Translocal orders of interest aggregation, policy-making and legal norms can either encourage or discourage recognition of collective interests, like those in environmental policy (cf. Calmfors & Drifill, 1988). Although local choices partly reflected contrasts between national political cultures, even local actors who pursued similar objectives in the two countries faced different sets of political interest organization, regulation and market conditions. Two such influences assured that, once in power, the Progressive coalition in Freiburg wielded official capacities and mobilized societal interests in ways that limited how far businesses and privileged groups could pursue environmental goods as luxury goods. Parallel infrastructures in Madison had the opposite effect. First, more extensive institutional means to exercise control over local markets for development gave the coalition in Freiburg greater capacities to pursue environmental ends through official policies. Legal mandates, financial resources, technical competence and organizational capabilities enhanced the leverage of the officials and activists who attained power. In Madison, faced with fewer official capacities, the coalition had to rely more on market actors and on localized neighborhood interests in environmental amenities. Second, the parties and corporatist forms of interest aggregation in Freiburg enabled environmental matters to be framed and pursued more as collective interests, and marginalized smaller, more particularistic groups. The nonpartisan, less organized structures of interest aggregation in Madison required the coalition to approach environmental questions more through particularistic representation of neighborhoods, businesses and households.

A full understanding of these orders and their consequences needs to take account of how these effects became compounded over time. Institutional differences fostered divergent paths of transformation in local political cultures that ultimately 'fed back' into local environmental policy (cf. Pierson, 1994). In Freiburg the more integrated patterns of settlement that the coalition maintained limited how far privileged residents and businesses could pursue special amenities for particular neighborhoods. To obtain environmental amenities for themselves, privileged citizens had to seek collective environmental goods. In Madison, segregated patterns of settlement reinforced the interests of privileged groups in a more exclusive, more divisible provision of environmental goods. We can see the consequences from a survey of the evolving coalitions and the conditions they faced.

Translocal Orders and Control Over Markets

National and Land-level systems of legal, organizational and financial resources gave the Progressive coalition in Freiburg greater capacities to control markets for land use. Relying on these and other resources, the coalition had not only provided more public environmental amenities but had contained spatial disparities. Despite the use the coalition in Madison made of more limited resources, public as well as private actors in the regime helped construct the selective and privately provided amenities of the West and East Sides. Eventually the residents of these privileged areas would themselves impose market and political constraints on the efforts of the Progressive coalition to provide public environmental goods.

The Progressive coalition in Freiburg drew on ample translocal resources. Backed up by constitutional qualifications to property rights, provisions of the elaborate German National Construction Law called on local officials to undertake such activities as to maintain the character of neighborhoods and landscapes (BbauG §34–35 (1994)). After 1975, environmental protection laws required officials to compensate for new development that intruded on natural areas or landscapes (NaturG §8 (1987)). Expansion of municipal boundaries in local government reforms at the beginning of the 1970s assured the city direct regulatory control over large portions of undeveloped land beyond the most urbanized area. As the staffing figures of Table 5.1 indicate, a larger local land use bureaucracy exercised control over markets. The national civil service and an extensive system of university training, professional certification and shared expertise among local land use officials helped foster high levels of technical competence and bureaucratic authority. Parapublic companies, integrated through nationally regulated systems of provision but entirely owned by the city, provided services from gas, electricity, water, sewer, transit and garbage collection to housing construction, urban development and financial markets. In further support of municipal initiatives, a standardized system of revenue distribution among Länder and local jurisdictions helped ensure financial resources for local administration. A national program to reconfigure the territorial boundaries of local government in the late 1960s and early 1970s facilitated annexations of outlying territories in order to extend municipal control over development. Enhanced through national frameworks at around the same time, regional planning reinforced and built upon the efforts of central cities. Constraints the law imposed on the private acquisition and development of land generally sent prices for new housing in the urban periphery higher than in the United States and much of Europe (Bundesministerium für Raumordnung, Bauwesen und Städtebau 1995). The resulting costs further discouraged developers and other private owners from challenging municipal agendas to manage growth.

Legal, financial, technical and organizational capabilities still depended in large measure on how the coalition and its representatives mobilized them. In Freiburg these actors had mobilized effectively around efforts both to provide high levels of environmental goods and to alleviate the disadvantages of working class neighborhoods. In Madison the coalition also drew on support from working class and comparatively disadvantaged neighborhoods, but in Freiburg local policies reshaped housing and transportation markets to favor these areas. Local initiatives of this sort in the German city outstripped parallel efforts among the eight mid-size French and German cities I studied (Sellers, forthcoming), as well as in metropolises like Hamburg (Dangschat, 1992) or Munich (Breckner & Schmals, 1992). As Freiburg acquired a reputation for environmental policy within Germany, and even beyond (e.g. Apel & Pharoach, 1995), city elites looked increasingly to public environmental goods to attract and maintain visitors, residents and businesses. The coalition planted more forested land. It secured unusually high usage of transit and bicycles. It extended pedestrian zones throughout the downtown. It established transportation routes, parks, and services that brought increasingly mixed residential populations to the western neighborhoods of working class high-rises. In approvals and sponsorship of new projects, it continued to assure housing for a mixture of different income levels. Along with aggressive use of other legal, financial and organizational resources, extensive ownership of local land aided pursuit of these ends. In the late 1960s, largely as a result of the extensive forests that the city owned, Freiburg controlled over half of the land within city boundaries. Ownership enabled the city to protect land that might otherwise be developed, to dedicate new land for development as a substitute for privately owned land, and ultimately to keep the cost of new private development high. By the early 1970s the city had also annexed a wide swath of surrounding villages (Ebnet, Günterstal, Hochdorf, Opfingen, Tiengen, Waltershofen). With this acquisition local officials secured direct regulatory control over a wide urban buffer of forest and open space.

The translocal resources available to political and administrative actors in Madison to control markets fell far short of the means available in Freiburg. In contrast with German constitutional provisions, U.S. constitutional property rights have long supplied a basis for legal challenges to local regulation and impinge on local regulatory authorities (Blaesser & Weinstein, 1989). U.S. developers as well as other landowners have seized on restrictions in state as well as national legal provisions to mount more frequent litigative challenges to local regulation than their German counterparts (Sellers, 1995). As a result of US rules themselves, as much as this threat of litigation, local political and administrative officials exercise less influence over trends in local development. More extensive fiscal encouragement to home ownership (Marcuse, 1982) enhanced market opportunities both for developers and for the middle and upper middle class homeowners whose houses the developers built. State-administered programs of peripheral highways like the Beltline around Madison added to these opportunities, with new physical infrastructure for peripheral neighborhoods. There was no national civil service at the local level to impose standards and consistent administrative practices as in German cities. Nor did local planning, housing and regulatory offices possess the prestige or administrative capacities of their German counterparts.[3] Slight supralocal regulation reinforced the role of local parapublic enterprises in markets related to development. The city controlled water, sewer and bus services within and beyond its borders, and the county controlled the local airport. But regulated private utilities, private municipal contractors and private financial institutions undertook many of activities that in Freiburg remained under exclusive municipal authority. In the central neighborhoods of Madison, a small but growing sector of nonprofit housing and social service organizations assumed some of the same functions as parapublic enterprises in Freiburg. This sector operated through grants and contracts that depended on more attenuated ties to local officials. The limited system of revenue sharing among localities in Wisconsin did far less than the national German system to bolster municipal finances. In the 1990s the city received 62% of its revenues from local taxes, compared with 22% from locally controlled taxes in Freiburg. Along with the comparative availability of undeveloped land in the U.S. Midwest, less extensive constraints on land use and development from higher levels of government generally favored private homeownership more than in Europe (Nivola, 1999).

By the 1990s, the Progressive coalition of Madison had drawn on much of the means available to it in pursuit of environmental goods. At times these efforts went so far as to secure more equitable provision of environmental amenities among neighborhoods. As early as 1973, the city had worked with a regional planning commission and zoning officials in the surrounding county to carry out a metropolitan plan that included open space corridors and a growth boundary similar to that imposed in Portland and a number of U.S. cities. Since 1977 city planners have employed extra-jurisdictional authorities to zone (93–94 Wis. Stats. 62.23 (7a)), alongside annexation powers to enforce a boundary on urban development (93–94 Wis. Stats. 66.021). The sanction of withholding city water and sewer services reinforced the growth boundary, and brought new development under municipal jurisdiction. These measures helped slow the loss of agricultural and forest land. By 1997, over a century of municipal purchases of land for parks and other purposes had brought just over 10% of land in the city under control of the local government. In the poorest section of South Madison, the city provided a community center, a recreation area, other parks and bicycle paths. Measures to preserve the viability of the downtown through new development and historic preservation limited flight to suburban areas. The local government itself constructed and administered the downtown pedestrian mall along State Street. Local officials also claimed much of the credit for bringing Monona Terrace, a long-disputed Convention Center and pedestrian area designed by Frank Lloyd Wright, to the downtown. The municipal-owned bus service helped link central neighborhoods to outlying shopping areas and offices.

At the same time, the neighborhoods of the West Side and to a lesser degree the East Side had consolidated the packages of environmental and other amenities already described. Private developers, businesses, consumers and officials acting independently from city authorities laid much of the foundation for these disparities. In promoting the residential neighborhood of Nakoma in the early decades of the twentieth century, local promoters held out the undisturbed natural environment of the area as the main selling point for life outside the urban center (Groy, 1982, p. 21). Additional advantages reinforced the preeminent position of such privileged areas. In the 1970s malls on the east and west sides surpassed the central business district as the main centers of retail services. The state assumed responsibility for the Beltline extension, and helped consolidate West Side suburbs with new office buildings of its own. Other public measures that remained partly separate from initiatives of the city government, like a technology park linked to the University of Wisconsin, preserved expanses of forest and open space alongside workplaces. With persistent demand for the housing and selective amenities of the periphery, middle- and upper-middle-class residents rewarded and reinforced the decisions of developers to expand there.

City officials contributed to the selective amenities of the outlying neighborhoods. Even under the Progressive coalition, these areas provided places for local political and economic elites to live, and helped attract and keep additional well-to-do and middle-class residents and businesses within the city. Throughout the postwar period, the city reaffirmed the position of the outlying neighborhoods with planning, zoning, approvals and participation in higher tier initiatives like the highway programs. On its own the city created municipal golf courses, parks and nature reserves that enhanced the advantages of these same areas. Schools built in these neighborhoods added to environmental amenities a set of educational institutions with the best reputations in the city system.

The residential interests that ultimately emerged in the periphery increasingly imposed political constraints that were not evident in Freiburg on expanding environmental amenities. From the 1970s onward, the middle-class and upper-middle-class constituencies in these peripheral areas emerged as the dominant political constituency of the city. Faced with the need to maintain electoral support among these shifting constituencies, as well as to carry out policies in the changing local economy, Soglin and other coalition leaders turned increasingly to strategies that limited overall levels of environmental goods. In several ways the resulting policies often reaffirmed or accentuated place-based disparities.

First, Soglin and his coalition turned to a limited 'fiscal populism' (Clark & Ferguson, 1983). His successful campaigns of the 1970s capitalized on his 'radical' image and university ties (*Capitol Times*, April 4, 1973, p. 4). From 1989 onwards, he placed as much emphasis on reducing property taxes as on land use issues (*Capital Times*, April 3, 1991, p. 4A; *Wisconsin State Journal*, April 5, 1995, p. 7A). Holding the line on demands for more municipal spending and taxation helped him broaden his appeal to majorities in most of the East and West Sides. Fiscal restraint acknowledged that residents who already enjoyed selective and privatized environmental amenities would give only limited support to broader environmental measures.

Second, policies to acquire additional parkland and to plan by means of individual neighborhood proceedings seldom redressed the systematic disparities of the city. Both of these policies divided up access to environmental amenities in ways that granted well-to-do neighborhoods as well as moderately disadvantaged areas new political opportunities to secure selective benefits to local residents. The bulk of the new parkland and new reserves of open space came from peripheral zones that were nearer to privileged neighborhoods. The neighborhood planning of the 1980s and 1990s began after the West and East sides gained a decisive place in the Progressive environmentalist coalition. In a process funded partly through the Federal Community Development Block Grant program, procedures for planning by small areas placed growing responsibility for securing environmental goods in the hands of neighborhoods. Representatives of neighborhoods assumed the initiative to propose new parks, open space, and other improvements in their residential areas. Despite the divisible benefits this planning brought to disadvantaged as well as privileged neighborhoods, the framing of amenities in this fashion obscured the wide disparities across neighborhoods. Both limits on the funds to be distributed and low participation on behalf of more amenities in places like the poor neighborhoods of South Madison (see below) hampered neighborhood planning. As a means of equity, this process of planning had at most helped mixed neighborhoods in the downtown to secure somewhat higher levels of amenities relative to the suburbs.

With the blessing and often the extensive participation of the city government, ad hoc business and institutional partnerships reinforced the advantages of outlying neighborhoods The projects that grew out of these arrangements furnished jobs and often shopping, services, and infrastructure to the residents of peripheral neighborhoods. The most significant new commercial and office developments of this sort emerged from the initiatives of individual firms, or institutions like the Wisconsin Department of Agriculture. City officials negotiated and ultimately approved conditions for these developments, and in some instances helped finance them (see below). During the late 1980s and early 1990s several large new business parks and commercial facilities on the outskirts of the city came about in this way (Wisconsin Community Development Institute 1995).

Peripheral residents thus joined local economic and political elites in mobilizing around local strategies that diverged increasingly from those of the German city. In curbing peripheral development, the coalition in Freiburg also helped forestall the emergence of outlying residential and business interests that increasingly dominated the Progressive coalition in Madison. Still, the largest local party in Freiburg, the CDU, relied heavily on support in similar outlying neighborhoods to those that increasingly dominated Madison. Yet even that party worked vigorously to expand public environmental amenities. To understand the reasons for this contrast requires a closer look at the relation of politics to the local civic realm.

Translocal Orders and Interest Representation

Interests in local policy stem partly from supralocal policies, local markets and residential patterns, but also from institutionalized patterns of political and civic representation within cities. In Freiburg the activists and constituencies of the Progressive coalition organized within a system of strong local party organizations, a highly organized municipal bureaucracy, and corporatist practices of economic and social representation. Relatively speaking, these structures insulated the coalition from other civic influences, mobilized civic organization around political aims, and integrated attention to distributive concerns with environmental policies. In Madison, the weakness of local electoral organizations, a less extensive and powerful local bureaucracy and the absence of corporatist organization, confronted the Progressive coalition with the disorganization that urban regime theorists have found in other U.S. settings (Stone, 1989; Ferman, 1996). As a result, the coalition depended more on the constant support of neighborhood associations and individual businesses to sustain its pursuit of environmentalist objectives. This dependence fostered more selective or privatized environmental policies and implementation than in Freiburg, and further skewed outcomes toward privileged neighborhoods and business interests.

In Freiburg, the civil service bureaucracy charged with planning land use, along with representatives of national political parties, dominated the political arena. Comparatively strong formal powers, like appointment through a separate, nonpartisan election, control over departments and appointments, and authority to preside and vote in the local council, reinforced the hand of Mayor Böhme. Proportional representation and city-wide council elections reinforced the largely stable positions of parties in policy-making. Elections held every five to seven years, with voting only for council members and the Mayor, compounded the limits on other actors to influence the local government. In the most significant procedural challenge to Mayor Böhme and the council over the course of the 1970s and 1980s, a referendum yielded a voting majority against a convention center that he had advocated. Even in this instance, after the referendum failed to garner the majority of all registered voters that was required to defeat the center, the council voted to carry out the Mayor's plans for construction. In this case, as throughout the 1970s and early 1980s, the Freiburg Greens, and environmental movements linked to them, employed challenges to the Council and the established parties to provoke growing support for environmental policy. In 1989, once the Greens assumed a leading role in a Left majority, the insulation that had helped shield policy from direct challenge worked increasingly in their favor.

The highly organized parties mobilized civic activity in ways that channeled it according to the agendas and formations of the political arena. Organizational and associational networks linked to party formations extended deep into various sectors of civil society. The SPD dominated a network of traditional neighborhood associations. The Greens and a breakaway SPD faction helped organize neighborhood initiatives against development. CDU members kept up contact with church-related associations like choirs and the Cathedral Restoration Society. FDP activists stressed memberships in local business associations (Sellers, 1994, pp. 120–123). As the Greens demonstrated, civic activists of any sort could rarely avoid the need to cultivate connections to some section of the community. The organizational connections between the party system and diverse parts of society fostered efforts to integrate the distributive concerns of various neighborhoods into environmental policy. SPD members owed their positions to a party that depended on the votes and activism of working class neighborhoods. The Greens depended in similar ways on concentrations of students and other younger voters. Strong party organization made it difficult to separate out developmental and environmental policies elsewhere in the city from the interests of these neighborhoods.

Alongside political organization, corporatist practices further regulated and constrained opportunities for participation. Planning formally incorporated the Chamber of Commerce and Industry, other business interests and activist groups like environmentalist associations. Simultaneously, this structure of opportunities confined the chances for both types of group to exert influence outside formal processes. Regular procedures gave neighborhood or business groups with interests in securing narrow, selective environmental amenities only limited chances to veto or secure leverage in opposition to broader environmentalist aims. Institutionalized representative processes of this sort seldom yielded major changes in projects. According to ratings by elites and activists, business and professional associations asserted 'considerable' influence on outcomes in only three of 34 local controversies (Sellers, 1994, p. 194). Although respondents attributed citizen or environmentalist groups 'considerable' or better influence in up to half of the controversies where these groups participated, other ratings showed these groups to have attained only 'slight' change to projects after the start of formal proceedings. In the face of similar constraints, developers often looked to party connections as a way of securing favorable treatment from local government. Although some contended that it made little difference for them, several prominent local builders openly cultivated party ties.

In Madison, in the absence of strong parties, a powerful bureaucracy, and formal corporatist procedures, the translocal infrastructures that shaped interest aggregation encouraged a politics of divisible and privatized goods. Despite formal executive-legislative relations that reinforced the position of the Mayor, the multiple agencies, commissions and other institutional actors responsible for aspects of land use fostered a largely open, incoherent organizational structure. Functions of local government not handled at the local level in Germany, like policing and public education, compounded the number of local organizational actors. Even the master plan for the city effectively depended on a wider array of more specific procedures for peripheral development, parks and open space, urban design and zoning. Formal procedures institutionalized the greater dependence of these and other processes of city politics on civic activity. Biennial elections, frequent referenda, public hearings, participatory procedures, and appointed or elected commissions opened official decision-making about environmental amenities and other issues at numerous points to more influence from wider publics and local veto groups. For instance, the statutory authorization for the City Planning Commission required three of seven members to be ordinary citizens and mandated public hearings with notice prior to decisions (93–94 Wis. Stats. 62.23). Analogous boards, operating under similar procedures, decided on zoning appeals and historical preservation. Opportunities to litigate on the basis of property rights also appear to have given wealthier neighborhoods and local businesses more means to challenge governmental initiatives than in Freiburg. In my comparison of another U.S. city with Freiburg (Sellers, 1995), both the U.S. developers and neighborhood groups had sued regularly and gained limited advantages from doing so. In Freiburg, by contrast, only environmentalist and citizen movements had employed litigation as a political means, and those groups generally lost in the courts.

Processes open to influence from both business groups and neighborhood-based organizations enhanced opportunities to secure particularistic benefits from policy. Individual businesses and institutions organized much new development on the periphery by means of ad hoc arrangements with the city government. The American Family Insurance Company had initiated the American Center with a proposal for its new world headquarters campus along Interstate 90–94. A private developer negotiated contributions from the city to his 17.2 acre Madison Corporate Center. In neither instance had city officials initiated the plan for the development or taken the leading role. In both, the city government had worked with the developers to secure approvals, infrastructure and even financing. Additional projects like High Crossing and the World Agriculture Center grew out of similar arrangements (Wisconsin Community Development Institute, 1995).

The Progressive coalition, organized through largely ad hoc coalitions of neighborhoods and other civic actors, depended increasingly on selectively provided amenities in neighborhoods. District-based council elections and neighborhood-based electoral organization further encouraged a politics of distinctive neighborhood interests. So did the nonprofit, neighborhood-based organizations for housing and social services that played a growing role in parts of downtown (cf. McGovern, 1997). Neighborhood planning also depended on capacities for activism evident in mixed downtown areas, which was lacking in the most disadvantaged areas (cf. Berry et al., 1993). Neighborhoods of committed middle-class activism, like those on the Isthmus, mobilized most effectively to take advantage of opportunities in neighborhood planning. In one of the mixed Isthmus neighborhoods, Marquette-Schenk-Atwood, some 30 local service organizations, four neighborhood associations, three 'neighborhood-based' organizations, two neighborhood newsletters and four separate subcommittees of individual local residents, participated (Madison 1994). Although the neighborhood already contained 11 beaches and other parks, the 40 page report from this procedure made 10 separate recommendations for new parkland or facilities, including bicycle paths and greenways.

In contrast, the most underprivileged neighborhoods suffered compounded disadvantages, despite the efforts of the city government. In 1980 one of the earliest neighborhood plans addressed the needs of the poorest neighborhoods of South Madison. Subsequent reports followed up these recommendations in 1988 and 1990 (South Madison Housing Task Force, 1988, South Madison Steering Committee, 1990). Participation by local residents lagged far behind levels in the Isthmus neighborhoods. Even in the first plan of 1980, which was backed by CDBG funding, local officials secured contributions from only a third as many local organizations (Madison, 1980). Although city officials prepared a lengthy report of some 55 pages as part of this process, its recommendations centered around such basic nuisances as noise, unsightliness, and bad drainage, and included only five new parks or facilities. In 1990, the Steering Committee that the Mayor appointed to take up the persistent problems of the area lacked federal support. This time the report listed only two associations based in the neighborhood, with a total of 19 individual participants aside from the Committee itself (South Madison Steering Committee, 1990). Composed overwhelmingly of officials plus business and institutional representatives, the Committee lacked the staff, the funding and the grassroots impetus to compensate for the weakness of neighborhood mobilization. The Committee's 15-page report, focused on employment, social opportunities for youth and safety in the neighborhoods, but paid little attention to environmental measures beyond transportation for local residents.

Green leaders in Freiburg expressed reservations about how the party system, the bureaucracy and corporatism there had restricted responsiveness to neighborhood concerns. But these institutions also facilitated pursuit of environmental objectives the Progressive coalition sought. These organizational forms empowered coalition leaders with respect to markets, limited and channeled the efforts of potential rent-seeking groups, and enabled the coalition to suppress or minimize the most particularistic and self-regarding neighborhood interests. If some actors in the Progressive coalition of Madison might have preferred similar strategies, the translocal context there fostered more reliance on neighborhood organization in elections, and on individual businesses and institutions in policy implementation. The resulting pattern of interest formation, representation and policy reinforced more modest, more selective environmental policies, and a greater dependence on private provision of environmental goods.

CONCLUSION: THE LESSONS FROM COMPARISON

The growing importance of environmental policy in urban politics necessitates analytic approaches that consider not only environmental objectives, but the ways those ends have combined with more established ones. Coalitions with similar Progressive identities in relation to the politics of a given national context may tie environmental objectives to social and economic ends in any of several ways. Even coalitions that follow Integrated policies in relation to their respective national contexts can look quite different from each other against the wider background of comparative international possibilities. Comparison of Freiburg and Madison suggests both the extent of these international divergences and how different paths have emerged. U.S. policies and organizational patterns, and accompanying residential interests, imposed different local conditions for environmentalist coalitions of the Left in Madison from those in many European cities. Even more favorable social and spatial conditions for Progressive environmentalism than in most U.S. cities could not prevent the emergence of Upscale orientations and policies, nor limitations on the extent of environmental measures.

Divergences between the trajectories of coalition politics and policy in the two cities highlight the crucial role that policies and institution-building at higher levels of government have played in the development of local coalitions that integrate environmental concerns with social and economic ones. Norms and institutions embedded at these higher levels empowered the coalition in Freiburg with policy mandates, financial resources and organizational tools to control markets, but left the coalition in Madison more dependent on support among private businesses and neighborhoods. At the same time, the more organized translocal systems of interest representation, bureaucracy and political parties in Freiburg enabled the environmentalist coalition to forestall resistance to the treatment of environmental concerns as collective goods. The institutionalized disorganization of interests in Madison compelled the environmentalist coalition to pursue the same goods through more divisible public means, or to leave provision to private markets. Both contrasts enabled businesses and privileged residents in Madison to gain more environmental goods for their own neighborhoods, and to oppose extension of those goods to the rest of the city.

Largely as a result of these contrasts, urban development evolved in directions that fostered distinct patterns of residential and development interests. The local political cultures, especially among the privileged constituencies who played important roles in both Progressive coalitions, reflected these divergent interests. In Freiburg, constituencies and activists accustomed to more integrated and centralized settlement continued to treat environmental amenities as public goods to be distributed equally. In Madison the growing, largely segregated constituencies of privileged residents developed interests in policies that provided environmental goods through more privatized, more selective and frequently more Upscale means.

In two cities with such similar political economies, that comprise such parallel places in wider systems of markets and space, most other influences offer less convincing reasons for the contrast in outcomes. In Madison, as in other U.S. cities, higher proportions of middle class and higher status residents may help to account for the greater bias of policy-making toward more privileged interests.[4] But a more concentrated residential pattern and more organized institutions of collective representation could just as well have encouraged these same privileged groups to pursue higher levels of urban environmental amenities in Madison than in Freiburg. U.S. policies and institutions not only foster dispersed, segregated neighborhoods of managerial and professional families, but encourage more self-regarding forms of neighborhood representation and business participation, which lead ultimately to fewer amenities. National and regional cultures cannot explain away these local effects (cf. Elazar, 1986; Lamont, 1992). If cultures at wider scales take their shape from the micro-level dispositions of the consumers and voters who make individual choices, then the propensities examined in these cities lie at the root of any such wider propensities.

Comparative studies of arenas like air and water pollution, and extensions to other cities, would help to fill out, strengthen and qualify these findings. A complete multilevel analysis would have to supplement such accounts with attention to the parallel logics that comparative analysts such as Weir (1996) have found in supralocal policies and politics. The growing recent emphasis on environmental justice in the U.S. environmental movement (Gottlieb, 1993) probably stems in part from reactions against the comparative environmental injustice that this comparison suggests is typical of the United States. So long as the institutional, spatial and cultural influences examined here continue to confront Progressive environmentalist coalitions in the United States, local policy toward both the environment and social equity will continue to fall short of German results.

NOTES

1. U.S. Authors like Jackson (1985) and Monkkonen (1988) have routinely cited northern and middle European countries like Germany as examples of effective control over sprawl. Recent studies also point to the greater success of these countries in such domains as the reduction of pollution and greenhouse gases and the implementation of recycling (Jahn, 1998; Scruggs, 1999). Urban governance plays an important role in many of these domains.
2. For rare exceptions see Savitch (1988); Apel & Pharoah (1995). For international comparative analysis from the perspective of national policymakers see Nivola (1999).
3. For instance, in a survey administered to local elites and activists in Freiburg and the U.S. city of New Haven, more Germans picked an administrative official as the most prestigious and best-paid on a list of five professionals, and saw the official as just as well trained as the others (Sellers, 1994, p. 236).
4. Although analysts have often noted the greater predominance of these groups in the United States (e.g. Esping-Andersen, 1993), the significance of this fact for urban politics remains largely unexamined.

REFERENCES

Altshuler, A. A., & Gómez-Ibáñez, J. (1993). *Regulation for Revenue*. Washington DC: Brookings Institution.
Amt für Statistik und Einwohnerwesen (1978). *Jahresheft 1977*. Freiburg im Breisgau.
Amt für Statistik und Einwohnerwesen (1988). *Jahresheft 1987*. Freiburg im Breisgau.
Apel, D., & Pharoach, T. (1995). *Transport Concepts in European Cities*. Aldershot: Ashgate.
Banfield, E., & Wilson, J. Q. (1963). *City Politics*. Cambridge: Harvard University Press.
Barnekov, T., Boyle, R., & Rich, D. (1989). *Privatism and Urban Policy in Britain and the United States*. Oxford: Oxford University Press.
Berger, R. (1995). *SPD und Grüne*. Opladen: Westdeutscher Verlag.
Berry, J. M., Portney, K. E., & Thomson, K. (1993). *The Rebirth of American Democracy*. Washington DC: Brookings Institution.
Breckner, I., & Schmals, K. M. (1992). München: Zwischen Isarbrücke und Luxus-Wohnung. In: H. Heinelt, & M. Mayer (Ed.), *Politik in Europäischen Städten*, (70–98). Basel: Birkhäuser.
Bridges, A. (1998). *Morning Glories*. Princeton: Princeton University Press.
Bullard, R. (1993). Anatomy of Environmental Racism and the Environmental Justice Movement. In: R. Bullard (Ed.), *Confronting Environmental Racism*, (15–39). Boston: South End Press.
Bundesministerium für Raumordnung, Bauwesen und Städtebau (1995). *Kostensenkung und Verringerung von Vorschriften im Wohnungsbau*. Bonn.
Calmfors, L., & Drifill, J. (1988, April). Bargaining Structure, Corporatism, and Macroeconomic Performance. *Economic Policy 6*, 14–61.
Clark, T. N., & Ferguson, L. C. (1983). *City Money*. New York: Columbia University Press.

Clark, T. N., & Goetz, E. G. (1994). The Antigrowth Machine: Can City Governments Control, Limit or Manage Growth? In: T. N. Clark (Ed.), *Urban Innovation,* (105–145). Thousand Oaks, California: Sage.
Clavel, P. (1986). *The Progressive City.* New Brunswick, New Jersey: Rutgers University Press.
Clavel, P., & Kleniewski, N. (1990). Space for Progressive Local Policy. In: T. Swanstrom & J. R. Logan (Eds.). *Beyond the City Limits,* (199–234). Philadelphia: Temple University Press.
Dangschat, J. S. (1992). Konzeption, Realität und Funktion 'Neuer Standortpolitik' – am Beispiel des 'Unternehmens Hamburg'. In: H. Heinelt, & M. Mayer (Eds.), *Politik in Europäischen Städten,* (29–48). Basel: Birkhäuser.
DeLeon, R. (1992). *Left Coast City.* Lawrence: University of Kansas Press.
Elazar, D. J. (1986). *Cities of the Prairie Revisited.* Lawrence: University of Kansas Press.
Esping-Andersen, G. (1993). *Changing Classes.* London: Sage.
Fainstein, S. S. (1993). *The City Builders.* Oxford: Blackwell.
Ferman, B. (1996). *Challenging the Growth Machine.* Lawrence: University of Kansas Press.
Freiburg im Breisgau (1995). *City Budget.*
Garber, J., & Imbroscio, D. (1996). The Myth of the North American City Reconsidered: Local Constitutional Regimes in Canada and the United States. *Urban Affairs Review 31,* 595–601.
Gottlieb, R. (1993). *Forcing the Spring.* Washington DC: Island Press.
Groy, J. (1982). Suburban Development in Madison. *Journal of Historic Madison 7,* 15–24.
Harding, A., Dawson, J., Evans, R., & Parkinson, M. (Eds.). (1994). *European Cities Towards 2000.* Manchester: Manchester University Press.
Jackson, K. (1985). *Crabgrass Frontier.* New York: Oxford University Press.
Jahn, D. (1998). Environmental Performance and Policy Regimes: Explaining Variations in 18 OECD-Countries. *Policy Sciences 31,* 107–131.
Jahn, T. (1996). Urban Ecology – Perspectives of Social-Ecological Urban Research. *Capitalism, Nature, Socialism 7,* 95–101.
John, P., & Cole, A. (1998). Urban Regimes and Local Governance in Britain and France: Policy Adoption and Coordination in Leeds and Lille. *Urban Affairs Review 33,* 382–405.
Judd, D., & Parkinson, M. (Eds.). (1990). *Leadership and Urban Regeneration.* Newbury Park, California: Sage.
Keating, M. (1991). *Comparative Urban Politics.* Aldershot: Edward Elgar.
Lamont, M. (1992). *Markets, Morals and Manners.* Chicago: University of Chicago Press.
Lauria, M. (Ed.). (1997). *Reconstructing Regime Theory: Regulating Urban Politics in a Global Economy.* London: Sage.
Le Galès, P. (1993). *Politique Urbaine et Développement Local.* Paris: L'Harmattan.
Le Galès, P. (1995). Du Gouvernement des Villes a la Gouvernance Urbaine. *Revue Francaise de Science Politique 45,* 57–95.
Leo, C. (1998). Regional Growth Management Regime: The Case of Portland, Oregon. *Journal of Urban Affairs 20,* 363–394.
Lewis, P. G. (1996). *Shaping Suburbia.* Princeton: Princeton University Press.
Liepietz, A. (1995). *Green Hopes.* Translated by Malcolm Slater. Cambridge: Polity.
Madison (1980). *South Madison Neighborhood Plan.*
Madison (1985). *A Land Use Plan for the City of Madison: Plan Report.*
Madison (1992). *City Budget.*
Madison (1994). *Marquette-Schenk-Atwood Neighborhood Plan.*
Madison (1995). *1995 Community Profile.*

Marcuse, P. (1982). Determinants of State Housing Policies: West Germany and the United States. In: N. I. Fainstein, & S. S. Fainstein (Eds.), *Urban Policy under Capitalism,* (83–115). Beverly Hills, California: Sage.

Martinez-Alliez, J. (1995). The Environment as a Luxury Good or Too Poor to be Green. *Économie Appliquée 48,* 215–230.

Massey, D. S., & Denton, N. A. (1993). *American Apartheid.* Cambridge, Massachusetts: Harvard University Press.

McGovern, S. (1997). Cultural Hegemony as an Impediment to Urban Protest Movements. *Journal of Urban Affairs 19,* 419–443.

Michelin Tire Company (1995). *Michelin Deutschland.* Karlsruhe: Michelin Reifenwerke.

Mollenhoff, D. (1982). *Madison: A History of the Formative Years.* Dubuque, Iowa: Kendall-Hunt.

Monkkonen, E. (1988). *America Becomes Urban.* Berkeley: University of California Press.

Nivola, P. (1999). *Laws of the Landscape.* Washington DC: Brookings Institution.

Obertreis, R. (1988). Ökotopia im Schwarzwald. *Baden-Württemberg,* Sonderheft 1: 64–65.

Page, E. (1991). *Localism and Centralism in Europe.* New York: Oxford University Press.

Page, E., & Goldsmith, M. (Eds.). (1987). *Central and Local Government Relations: A Comparative Analysis of West European Unitary States.* London: Sage.

Pierson, P. (1993). When Effect Becomes Cause: Policy Feedback and Political Change. *World Politics 45,* 595–628.

Prigge, W., & Lieser, P. (1992). Metropole Frankfurt: Keine Metro, Aber Polarisierung. In: H. Heinelt & M. Mayer (Eds.), *Politik in Europäischen Städten,* 49–69. Basel: Birkhäuser.

Ronneberger, K. (1994). Zitadellenökonomie und Soziale Transformation der Stadt. In: P. Nolle, W. Prigge and K. Ronneberger (Eds.), *Stadt-Welt,* (180–197). Frankfurt am Main: Campus.

Rosdil, D. (1991). The Context of Radical Populism in US Cities: A Comparative Analysis. *Journal of Urban Affairs 13,* 77–96.

Savitch, H. V. (1988). *Post-Industrial Cities.* Princeton: Princeton University Press.

Scruggs, L. (1999). Institutions and Environmental Performance in Seventeen Western Democracies. *British Journal of Political Science 29,* 11–31.

Sellers, J. M. (1994). *Public Authority and the Politics of Metropolitan Land in Three Societies.* Unpublished Ph.D. Dissertation, Yale University.

Sellers, J. M. (1995). Litigation as a Local Political Resource. *Law and Society Review 29,* 475–516.

Sellers, J. M. (1998). Place, Post-industrial Change and the New Left. *European Journal of Political Research 33,* 187–217.

Sellers, J. M. (1999). Public Goods and the Politics of Segregation. *Journal of Urban Affairs 21,* 237–262.

Sellers, J. M. (forthcoming). *Governing from Below: Urban Regions and the Global Economy.* Cambridge: Cambridge University Press.

Silver, M. L., & Melkonian, M. (1995). *Contested Terrain: Power, Politics and Participation in Suburbia.* Westport, Connecticut: Greenwood.

South Madison Housing Task Force (1988). *Housing Task Force Report.* Madison.

South Madison Steering Committee (1990). *Short-Range Recommendations for South Madison.* Madison.

Stadtplanungsamt Freiburg (1980). *Flächennutzungsplan Freiburg – Erläuterungsbericht.* Freiburg im Breisgau

Stone, C. N. (1989). *Regime Politics.* Lawrence: University of Kansas Press.

Stone, C. N. (1993). Urban Regimes and the Capacity to Govern. *Journal of Urban Affairs 15,* 1–28.

Strom, E. (1996). In Search of the Growth Coalition: American Urban Theories and the Redevelopment of Berlin. *Urban Affairs Review 31,* 455–481.

Tressel, R. (1990a). Der wachsende Pendlerstrom. In: Amt für Statistik und Einwohnerwesen (Eds.), *Stadtverkehr Wohin?*, (7–54). Freiburg im Breisgau.
Tressel, R. (1990b). Verkehrsverhalten von Studenten. In: Amt für Statistik und Einwohnerwesen (Eds.), *Stadtverkehr Wohin?*, (55–57). Freiburg im Breisgau.
United States Bureau of the Census 1993). *1990 Census of Population and Housing: Population and Housing Characteristics for Census Tracts and Block Numbering Areas: Madison, Wisconsin MSA*. Washington DC: Government Printing Office.
Weir, M. (1996). Poverty, Social Rights and the Politics of Place. In: Pierson & S. Leibfried (Eds.), *European Social Policy,* (329–354). Washington DC: Brookings Institution.
Wisconsin Community Development Institute. (1995). *Community Development Activities in Madison.* Madison.
Wolman, H., & Goldsmith, M. (1992). *Urban Politics and Policy.* Oxford: Blackwell.
Zeuner, B., & Wischermann, J. (1995). *Rot-Grün in den Kommunen.* Opladen: Leske.

6. THE RESPONSIVENESS OF LOCAL COUNCILORS TO CITIZEN PREFERENCES IN STUTTGART: THE CASE OF SPORTS POLICIES

Melanie Walter

In *Polyarchy*, Dahl (1971, p. 1) devotes one chapter to the control of representatives and requires ". . . that a key characteristic of a democracy is the continuing responsiveness of the government to the preferences of its citizens, considered as political equals." Responsiveness in this context means sensitivity to citizen wishes and attendant parliamentary behavior (see Pennock, 1952; Eulau et al. 1959; Pitkin, 1967). For Lijphart (1984, p. 1) also, responsiveness is essential to democracy, which: ". . . may be defined not only as government by the people but also, in President Abraham Lincoln's famous formulation, as government for the people – that is, government in accordance with the people's preferences. An ideal democratic government would be one whose actions were always in perfect correspondence with the preferences of all its citizens." It follows that an important question for the empirical theory of democracy is: "To what degree do democratic systems accomplish this goal?" (Cnudde & Neubauer, 1969, p. 6). Viewed in this light, the main question is how far representatives take account of the public's interests in reaching policy decisions? Since the 1970s some researchers have examined this relationship at the national level (e.g. Backstrom, 1977; Monroe, 1979; Brooks, 1985, 1987, 1990; Jackson, 1992). However, there has been little research at the micro-level exploring relationships between constituency preferences and the roll-call behavior of representatives.

Early work on these linkages owes much to Miller & Stokes (1963). They identified three types of representation: the *instructed delegate*, who represents the interests of her/his constituency in policy-making; the trustee, who follows her/his own conscience; and, the *responsible party* representative, who is largely driven by her/his own party's position. For the U.S. political system Miller & Stokes found a mixture of these representation models, with the orientation of politicians differing across policy issues. In this context Miller & Stokes established the concept of *congruence*; which is the degree of policy agreement between voters and the roll-call behavior of incumbents. This embodies the relationship between constituency preferences and its representative's attitude, as well as between constituency preferences and representative perceptions of constituency preferences (see Cnudde & McCrone, 1966). Broadly speaking, a constituency can influence the policy actions of its representative in two ways. First, it can choose a representative who shares its views, so even if a representative follows her/his own convictions these represent constituents' wishes. Alternatively, a representative can follow her/his perceptions of constituency attitudes. Potentially, these two are inter-related, as representatives probably see their districts as having similar opinions to their own or over time tend to bring their opinions more into line with their constituents.

Although some studies in Germany have considered links between the preferences of politicians and the public, only a few exist at the local level. This is particularly regrettable because democratic theory ascribes a closer relationship between elected officials and constituencies at the local level (e.g. Sharpe, 1970). This study seeks to address this gap through an analysis of citizen and councilor attitudes toward local sports policies in Stuttgart.

APPLYING THE MILLER-STOKES MODEL

To start we should recognize that the transposition of the Miller-Stokes design to a European context is not problem free. The main reason is that this model was developed to analyze the U.S. political system. The model is less applicable to parliamentary systems, where the government's stability depends on party support, with a greater requirement for representatives to vote in accordance with party policy. Strong party cohesion is maintained by the recruitment process for representatives which is dominated by parties. Put simply, compared with the USA, in much of Europe it is the party, rather than the local constituency, that is the stronger reference point for representatives.

Added to this, there are particular difficulties in assessing linkages between politician and voter preferences. For one, in Europe it is rare for roll-calls to be recorded (or at least be available for researchers), especially at the local level. This usually means it is necessary to identify representative's attitudes, as a proxy for actual behavior. Given that elections require voters to evaluate representatives across a range of policies, the only single issue on which constituency preferences and representative priorities can be directly related is through referenda. However, the value of referenda data is limited in Europe.[1] Using voting figures more widely is further restricted as the concept of the median-voter has limited application (Downs, 1957). This is because two party systems are in the minority. Again, the analyst is pushed toward identifying stated (citizen) preferences, rather than calling on measures of actual behavior. In operationalizing the linkage between citizen and representative preferences there are two main evaluative measures:

- Is public opinion identical with representatives' perceptions?
- How far are representative preferences coincident with their evaluations of public preferences?

A critical issue here is what is meant by 'the public'? Quite feasibly representatives believe they are responsive to public opinion but link this to the opinions of their own party's supporters. In this case, they are responsive, but not to the whole public. In order to evaluate this question, the analysis undertaken here compares representatives' views with the mean average score for public opinion and with the modal view amongst the public, as well as using the same measures against those who voted for a representative's party. The aim is to identify which representatives' views most closely match, as an indication of which unit representatives have in mind when deciding public policy.

To explore these relationships data are drawn from two questionnaire surveys. The first is a random sample survey of public opinion with 476 respondents.[2] The second is a complete survey of the 60 members of the local council in Stuttgart; of whom 43 councilors answered the questionnaire. These surveys were executed using the same standardized questions. Politicians were also asked about their perception of public opinion and their own responsiveness to that opinion. In this chapter sports policies are focused upon. This policy field is an important one in Stuttgart, for large sums of money have been spent on sports over the last decade (e.g. international events and large sports complexes). When the questionnaire surveys were conducted (the summer of 1993), spending on sports was contentious, as there was less money to spend on all public services. Allied with this, dispute existed over a trend in spending that lessened inputs into (often commercial) traditional sports events in favour of social or health related sports activities (Table 6.1). Central to this change has been

Table 6.1 Traditional and Modern Sport Policies in Stuttgart

Traditional sport policies	Modern sport policies
• Support for top-class sports events • Support of large scale events • Organization of large-scale leisure time events • Support for club sport competitions • Support of popular and leisure sports in clubs	• Support for healthy sports/sports for old people in clubs • Support for popular/leisure sports by free associations of citizens • Support for healthy sports/sports for old people, in free associations of citizens • Organization of popular/leisure sports

a reorientation in sports organization. Traditional sports activities happened largely in sports clubs, but newer sports activities are more prone to involve free associations of town citizens. Additionally, differences arise in the content of sports. Traditional sports events emphasize competition, whereas modern sports events stress social and health goals.

PERCEPTION OF PUBLIC PREFERENCES

As Pennock (1952, p. 791) makes clear, accurate perception of the public's preferences can be hard for representatives: "Often there are many wills, many opinions, but neither a consensus nor even a majority." The normative ideal of one public and one common interest is rare in the empirical world. Yet representatives do seem to be conscious of the heterogeneity of public opinion (Kurer, 1979). Indeed, according to Miller & Stokes (1963), the path from constituent preferences to representative perceptions (of those preferences), and then onto representative roll-call behavior, is a key element in representatives' behavior. Following this line, some studies have used representative perceptions as surrogates for 'real' public opinion (Kessel, 1964; Kingdon, 1973; Gabriel et al. 1993a). The survey data used here offers a chance to check the assumption that public opinion is reflected in representative behavior and attitudes.

In assessing this linkage, nine different sporting activities have been considered. These are:

- Top-class sport (TCS)
- Competitive club sports (CCS)
- Popular/leisure sports in clubs (PLC)

- Support for major sports events (SME)
- Large-scale leisure events (LLE)
- Popular/leisure sports events, open to all citizens (PLG)
- Health related sport for older people in clubs (HOC)
- Popular/leisure events in town (PLT)
- Health related sport, open to all older citizens (HOG)

For each of these categories, citizens were asked whether the money spent on these activities should be reduced significantly (scoring 1), reduced (2), kept at the same level (3), increased (4) or increased significantly (5). Representatives were asked what they thought citizen views on these points were and what their own views were. Computations were made of modal responses and mean average scores, using the scoring numbers listed above. In the main these results show that representatives' views on citizen opinions were fairly accurate at an aggregate level, with the highest level of misperception being for top-class sport, for which representatives were inclined to see citizens wanting larger reductions in spending than citizens themselves revealed. For six of the nine activities representatives consistently under-estimated the willingness of citizens to spend more on sport (for one of the nine items, the mean values for citizen views and representatives perceptions were identical). The two cases for which representatives evaluated citizens as wanting higher spending than the citizenry itself expressed were modern sports activities (PLG and HOC). Set against this, in every case, representatives preferred to see less spending than the general public. Put another way, for all activities except top-class sport, representatives' spending preferences were for fewer resources to be committed to sports than they believed the public wanted. Expressed in terms of mean average scores, the values recorded across all nine activities were 3.21 for the general public, 2.97 for representative views on what the public wanted, and 2.66 for what representatives themselves preferred.

It is reasonable to assume that incumbents better perceive preferences amongst their own party's supporters. In Stuttgart six political parties constitute the local council, with response levels for four of these being of sufficient magnitude to make valid comparisons of representatives and voters (CDU, SPD, FDP and GRÜNEN; Table 6.2). For sports policies Christian Democrat (CDU) representatives reveal almost perfect congruence between party voter preferences and politician perceptions of those preferences over the nine activities. For the other parties, the mean average deviation in citizen scores and politicians' evaluations of the same (0.30 to 0.36) is greater for own party supporters than for the population as a whole (0.24). The assumption of better perception of own party voters is denied for the most parties.

Table 6.2 Deviation of Representative from Public Spending Preferences by Party

Party	Public Preferences (PP)		Representative perceptions (RP)		PP-RP*
	Mean	N	Mean	N	Deviation
CDU	3.15	109	3.16	14	−0.01
SPD	3.23	84	2.87	11	0.36
FDP/DVP	3.04	33	2.69	6	0.35
GRÜNEN	3.04	59	3.34	3	−0.30

NOTE: A positive value indicates politician under-estimation of public opinion − perceiving less support for spending than exists. A negative value indicates over-estimation of public opinion − perceiving more support for spending than exists.

ASSOCIATIONS WITH REPRESENTATIVE PERCEPTIONS

Because of the difference between real and perceived public opinion, we can pose the question: what supports the dissonance recorded? The literature shows several factors influencing politicians' views of public opinion. In Table 6.3, the results of analyses exploring some of these factors are shown. Of the factors examined, six relate to personal attributes, like age or length of incumbency, while two relate to the context in which representative perceptions are made (the homogeneity of public opinion and issue importance).[3]

First examining personal factors, comparison of deviations between citizen views and politicians' images of those views reveal that older politicians (aged 60–78), and those with a long incumbency period (5–7 terms), have more accurate views of citizen preferences (confirming the hypothesis of Clausen, 1977). Clausen also suggests that politicians with a specialist interest can be expected to know more about public preferences in the relevant policy field (e.g. if they are members of the sports committee of the local council). This conclusion is also supported. Confirmation also obtains for a strong link between membership (as opposed to committee membership) of sports organizations, which was assessed here against the 360 sports organizations in Stuttgart, and the accuracy of representatives' perceptions of citizen views. However, relations with political party disposition produce no clear message. According to Dalton (1985), mass parties like the CDU and SPD should perceive majority public opinion better than small ideological parties, like the FDP or GRÜNEN, which are more prone to focus attention on their own party's voters. However, for

Table 6.3. Links with Differential Representative Perceptions

	Accuracy of representative perceptions		
Factors	Below average	Average	Above average
Age	31–59 years		60–78 years
Incumbency time	1–4 terms		5–7 terms
Sex	Women	Men	
Special interest	Not a sports committee member		Member of a sports committee
Party	SPD, FDP/ DVP, REP, FWV		CDU, GRÜNEN
Sport organisation	Not member		Member
Issue importance	Important		Less important
Homogeneity of public opinion	More heterogeneous		More homogeneous

sports policies in Stuttgart, Green politicians (GRÜNEN) perceive public opinion as well as the mass CDU and better than the SPD.

In terms of the contextual factors surrounding representatives' views, Clausen (1977) emphasizes that issue importance is critical. For more important issues, the expectation is that representatives are less inclined to ignore citizen preferences, owing to heightened public attention on the issue. However, it appears that for sports politics in Stuttgart, issue importance has a slight bearing on the accuracy of representatives' perceptions (Table 6.3). Possibly this arises because the citizen scores for the importance of various sports policies were similar (2.85–3.69), so politicians have little motivation to establish a finely-tuned evaluation of citizen values. Aiding representatives in this regard, when the standard deviation of citizen spending preferences was lower, so greater homogeneity existed in public views, politicians were inclined to anticipate public views with more accuracy.

At this point we have some answers about politicians' perceptions of 'real' public opinion. Perceptions appear to be more adequate for spending questions on modern sport policies, than with traditional sports activities, even if there is a tendency to under-estimate citizen willingness to spend on sport. Yet perception accuracy is not simply related to activity type but also to age, incumbency, special interests, organization membership and public opinion homogeneity.

PERCEIVED PUBLIC OPINION AND REPRESENTATIVE ATTITUDES

So far representative images of citizen preferences have been compared with citizen expressions of their own preferences. This only represents one link in the chain from citizen values to policy implementation. A further element is the relationship between representative perceptions of citizen views and representatives' personal attitudes. This is the responsiveness that is uppermost in politicians' minds, and for citizens it is important to see their views reflected in representatives' preferences. A significant issue here is that representatives might want to be responsive to the public, but misperceive public preferences.

To analyze this issue a mean comparison method is used. If we plot values for councilor perceptions of public preferences (x-axis) and representatives' own preferences (y-axis), then congruence would result if the mean average scores for each sports activity were the same for councilors' perceptions and councilors' own views. Deviations from a perfect match would reflect dissimilarities between the two. This relationship can be expressed by a linear regression formula, in which 'perfect' congruence would see the intercept [a] coefficient equal to 0, the slope [b] coefficient equal to 1, and the R^2 coefficient at 1. When this is done for the Stuttgart data, a nearly perfect relationship is found, with the R^2 coefficient at 0.92. With the regression slope coefficient at 0.70, the results show that councilors' own preferences tend to be less accepting of increased spending on sports events than they believe the general public are. The exception to note is traditional sports activities, for which councilors were more supportive of spending than they believed the public to be.

REAL PUBLIC OPINION AND REPRESENTATIVE ATTITUDES

There is a close relationship between councilors' own values and what they believe are the values of the citizenry. This might make local politicians feel comfortable proclaiming their responsive credentials, but these will carry less weight with the general public if councilors misperceive citizen preferences. The question that remains then is whether representative preferences reflect citizen priorities. In this regard we can inquire about the responses of individual councilors or of the council as a whole, just as responsiveness can be evaluated with regard to the whole council or party groups. Considering these possibilities, Weissberg (1978) distinguishes between collective and dyadic representation:

- At the *collective level* two comparisons can be made. The first is congruence between *mean* public opinion and the *mean* preference of all local councilors (i.e. of the council as a whole). The second element at the collective level is congruence between the *mean* preference of a party's voters with the *mean* preference of local councilors from the same party. This second

element focuses on 'responsible party model' ideals, in which politicians mainly represent the interests of party voters.
- At the *dyadic level* the reference point is not *mean* public opinion or the mean preference of party voters but the individual voter. Here the aim is to evaluate how far representative values conform with those of voters who make up their constituency (this approach recognizes diversity of views amongst voters in a constituency rather than focusing on the mean average value for a constituency's voters).

Collective Responsiveness

In terms of mean average scores for the two variables, there is a high degree of congruence between public and representative attitudes over sport issues. This finding is confirmed by Brettschneider & Walter (1996), who analyzed responsiveness for 13 other local issues in Stuttgart. Yet this conclusion is unusual in a German context. Thus, in Arzberger's (1980, p. 147) investigation of policy responsiveness in Frankfurt, Aalen, Coburg, Using and Hadamar, the conclusion reached was that: "... the politicians' preferences ... rarely show a similarity ... to mean public preferences." That said, for specific issues, there is considerable deviation in views between citizens and the council in Stuttgart. For one, in a representative survey of 1,255 citizens undertaken by the Department of Sports Science at the University of Stuttgart, it was found that 80% of the public supported sports organization via free associations of citizens (Wieland & Rütten, 1991). The city council and its sports office prefer traditional organization programmes that use sport clubs. As a consequence the council and its sports office react slowly to public opinion. The same situation occurs for popular/leisure sports. Overall, however, members of the city council have a good estimation of citizen attitudes. Deviations in their responsiveness arise from an under-estimation of people's preferences. The average citizen would spend more on local sports than politicians.

Of course, congruence in attitudes does not indicate congruence in behavior. Yet more than one-third of local politicians consider that they always or most of the time consider public opinion in reaching policy decisions. A further one-third claim to take the public's views into account much of the time, with only 22% rarely or never reacting to public preferences. In contrast to these results for the whole council, responsiveness for single parties showed a lesser congruence. The mean correlation coefficient indicates a positive connection between citizen and councilor preferences within parties ($R^2 = 0.10$), but this link is weaker than for the whole council ($R^2 = .87$). It follows that assessments of councilor responsiveness partly depend upon the population units against which comparison is made.

Dyadic Responsiveness

In the tradition of Miller & Stokes (1963), many researchers have discussed the problem of how best to measure the liberal democratic concepts of citizen equality, neutrality toward alternatives and popular sovereignty (e.g. Blalock, 1967; Achen 1977). In this context, Achen has offered three measures. The first is *responsiveness*. This derives from the liberal doctrine of popular sovereignty, which proclaims that what the people decide must influence the outcome (Achen 1978, p. 490). "Conservative districts should have legislators with right-wing views, liberal districts should be represented by left-wingers." The question is, if two constituencies differ in their views, what is the difference in view of their representatives? The concept of responsiveness is assessed here using unstandardized regression coefficients when constituent preferences are used to 'explain' that of their representative. This measure not only presumes that politicians should hold opinions similar to their constituencies, but defines this pairing in relational terms. "That is, [the] 'a' [regression coefficient] is the expected position of the representative when the constituency opinion is zero, while 'b' is the expected change in the representative's opinion as constituency opinion changes by one unit" (Achen, 1978, p. 490; Powell & Powell, 1978). Dyadic responsiveness occurs when the views of voters in a constituency (which can then be aggregated across constituencies) predict the opinion of the constituency's representative (or in aggregate representatives across the city). The intercept 'a' and the slope 'b' (unstandardized regression coefficients) describe the pattern of linkage. It follows that slopes and intercepts in Table 6.4 can be interpreted in the following way: slopes of 1 indicate that representatives' views change in tandem with those of their constituents, lower or higher slopes indicate a more sluggish or heightened responsiveness, respectively. The intercepts ['a'] in this Table are the expected opinion of a representative when mean opinion in her/his constituency is 0. In this context, the 'a' value is an indicator of bias in politicians' 'core' responses, compared with those of their constituents. Therefore, a relationship in which 'a' is equal to zero and 'b' is equal to 1 represents ideal dyadic responsiveness. In this case representative attitudes are predicted exactly by the opinions of the public.

The concept of *proximity* focuses more on the *absolute* distance between the position of single voters and the whole council. That means the aggregate distance of all voters from all representatives.[4] This concept of representativeness is related to the norm of equality in liberal representation theory. If people have opinions and a representative's function is to reproduce them, then everyone's voice should count equally (Achen, 1978, p. 481). The closer the agreement on policy of the individual voter and the councilor the more representative is the councilor (for that voter). It is expected that the homogeneity of public preferences inclines such distances to be smaller, since accurate councilor perception of citizen views is easier if citizens do not have a diversity of views on a specific policy question. But if public opinion falls into two extreme positions, it is difficult for the council to be responsive to modal public opinion (Achen, 1978). On this account, Achen offered the measure of *centerism*. This focuses on the distance between council and citizen views after heterogeneity in public opinion is taken into account. Centerism is based on the simple calculation of the *proximity* score minus the variance of public preferences.[5] The lower the *centerism* score, the more the council is at the center of voter opinion. "A definition of representativeness based on efficiency or centrality within a constituency reflects the liberal democratic norm of fairness or neutrality toward opinions. No ideological view is to be given special treatment; all compete for votes on equal terms, and only voter preferences determine the representative's views" (Achen 1978, p. 488).

Table 6.4 presents the computations made for these measures. In this Table, the nine sport categories are ordered hierarchical by the size of the correlation coefficient. First we can see a similar ordering of values for Pearson's 'r' and the *responsiveness* score of Achen. These three measures indicate relatively high agreement for support for top-class sport, support for large-scale events and for the organization of large leisure events. There is less congruence in support for competitive club sports and support for popular/leisure sport in clubs. Most noteworthy are the relatively low values for the *responsiveness* measures. The preferences of the council and of the citizenry are in the same direction but not at the same level. Pearson's r (0.16) indicates that public preferences only account for 2.5% of variance in politicians' preferences over all nine issues. In terms of responsiveness, the council hardly accentuates voter opinions on sport policies. There is a notable bias in politicians' views (a = 2.67) and the regression slope (b = 0.08) is the opposite of the theoretical ideal (a = zero and b = 1).

The degree of absolute agreement as measured by the *proximity* and *centerism* scores is higher than the dyadic responsiveness as measured by *responsiveness* or Pearson's 'r'. Note that there is no proximity score higher than 1.54 (although the maximum score is 16). This means that councilors – considered in absolute terms – are near the single voter. Once variance in public preferences is taken into account the differences decrease immediately. The high degree of responsiveness at the collective level (md-scores) is thereby verified by measures of proximity and centerism at the dyadic level. This suggests that Achen and Blalock's criticisms of using correlation and regression coefficients as measures of responsiveness require further investigation. These measures appear to provide lower estimates of

Table 6.4 Measures of Dyadic Responsiveness

Sport-policies	Measures of dyadic relationship single constituent – parliament						Collective level
	Pearson's r	Regression R^2	Responsiveness A	b	Proximity[1] R	Centerism C	MD^2
TCS	0.32	0.10	1.74	0.15	1.35	0.26	0.09
SME	0.28	0.08	2.04	0.11	1.54	0.25	0.51
LLE	0.25	0.06	2.27	0.10	1.50	0.36	0.60
PLG	0.20	0.04	3.08	0.12	1.33	0.01	0.12
HOC	0.13	0.02	3.55	0.006	0.77	0.00	0.05
PLT	0.12	0.02	2.91	0.09	1.33	0.44	0.67
HOG	0.10	0.01	3.11	0.07	1.47	0.54	0.73
CCS	–0.01	0.00	2.86	–0.00	1.08	0.24	0.49
PLC	–0.01	0.00	2.51	–0.01	1.01	0.27	0.52
∅	0.16	0.04	2.67	0.08	1.26	0.26	0.42

1. TCS = Top-class sport
2. CCS = Competitive club sports
3. PLC = Popular/leisure sport in clubs
4. SME = support for major events
5. LLE = Large-scale leisure events
6. PLG = popular/leisure sport in free associations
7. HOC = Health sport/sport for old people in clubs
8. PLT = Popular/leisure sport in town
9. HOG = Health sport/sport for old people in free associations

1 The proximity score is the mean deviation of the council position from the single constituent's position. The highest score is 16 and the lowest 0.
2 Md measures the difference between the mean public opinion and mean council preferences. Because preference scales reach from 1 to 5, the highest md-score is 4 and the lowest 0. Positive values indicate that the public what to spent more money than the politicians.

responsiveness than other measures. Further doubt on the validity of using correlation and regression as measures of responsiveness arise from the assumed linear relationship between dependent and explanatory variables. As McCrone & Stone (1986, p. 958) report, this assumption is questionable.

Interestingly, when the same calculations are made for representatives of a specific party and voters for that party, except for the Social Democratic Party (SPD), representatives reveal a poorer ability to identify their own supporters' views than the local council as a whole represents the views of all voters. In general, if representatives seek to pursue policies they think will score well with their own supporters, they are likely to anticipate wrongly citizen wishes by more than if they seek to address what they see as the public's wishes at large.

CONCLUSION

This chapter has analyzed the relationship between representatives and citizens in the local political system of Stuttgart. In doing so, it recognizes that the most appropriate measure of 'responsiveness' is not a straightforward issue. There are questions of how well politicians perceive 'real' public opinion, of whether they respond to perceived citizen views or place higher priority on their own views, and then how far actual policies are allied to 'real' (or perceived) citizen preferences (as opposed to politicians' wishes). In exploring some of these relationships, this chapter has reached five main conclusions:

- In broad terms, local politicians better perceive mean average public opinion than opinions of their own party's supporters (although not explored here, this result also applies to voters in their own municipal district). The only exception to this is CDU politicians.
- Representative perceptions of public opinion are not equivalent to 'real' public opinion, although the deviation of one from the other is quite low. That said, the relationship varies across policy issues. For traditional sport policies, representatives are less in tune with public opinion than they are for modern sport policies. For traditional sports activities, politicians seem to project their own preferences onto the public. It follows that the use of representative perceptions as a surrogate for 'real' public opinion is only adequate in some cases.
- Representative perceptions of public opinion are closer to that of citizens themselves the older the representative (viz. if 60–78 years), the longer s/he has held office (viz. incumbency of 5–7 terms), if the representative is a member of the sport committee on sport politics on the local council and/or a sports organization, and if there is greater homogeneity in public opinion. The sex of the representative, whether or not s/he is a member of a mass party and the importance attached to a particular policy issue, are not key elements in determining the accuracy of representative perceptions.
- The degree of responsiveness is different for perceived and real public opinion. Councilors most evidently project their own perceptions of public opinion, but the extent of responsiveness to both perceived and 'real' public opinion is in fact high. Taken together the council is more responsive to citizen preferences than representatives of single parties. Considering the reference point of responsiveness, the representation of mean public opinion is better than that of the single voter. That means the collective level of analysis shows better results than the dyadic level.
- Classical measures of representative responsiveness, that rely on correlation and regression coefficients, under-estimate responsiveness. Achen's (1977, 1978) two distance measures, which refer to absolute agreement between representatives and constituencies, indicated a better theoretical content and superior empirical numbers for dyadic correspondence.

One objection to these conclusions is that they are based on just one town, at one time point and for a single policy field. This points to the need for further research in different towns across a variety of policies (Gabriel et al. 1993a). For any following research, unsolved questions over adequate methods of measuring responsiveness need to be addressed. For example, in this paper the dependent variable has been politicians, but there is the question of how far views of administration officers, or of economic, organizational or culture elites, should be taken into account (see Hoffman-Lange, 1991, p. 280). All these people influence the policy-making process and the extent to which politicians take account of these reference groups might vary across policy fields, as well as depending upon local power structures.

It should also be acknowledged that policy responsiveness is a dynamic process, as well as being influenced by national or regional political discourses (see Brettschneider, 1995). On the one hand, congruence can result from politicians reacting to public opinion. On the other hand, politicians' behavior or attitudes can influence public preferences: a new policy can change public opinion, just as politicians can create public opinion by information and/or manipulation (Brettschneider, 1996).

Finally it is possible that the structure of a political system influences responsiveness. Hence, to show the characteristics of a political system that are important for the congruence of policy with public opinion, it is necessary to conduct comparative research. This could involve international comparisons, as well as comparisons of different levels in a political system (national, regional, local) or even different institutions (legislative, executive, judicial). Is responsiveness higher in presidential systems than in parliamentary ones, where you have to consider parties? Does federalism or unitary government show higher responsiveness? There is still a long way to go in understanding responsiveness issues, and comparative research is just starting.

NOTES

1. In Belgium, the Netherlands and the U.K., referenda are not part of the constitution. In Denmark, France, Ireland, Italy and Spain a referendum has to concern changing the constitution. In Germany and Italy questions of territorial change can be put to a referendum. In Greece, Portugal and Spain referenda are largely confined to addressing major national debates. In some countries a large number of voters have to support a referendum taking place. For instance, in Italy and Spain at least 500,000 citizens have to call for one. In other countries certain issues are explicitly excluded from a public vote, as with financial questions in Greece, Ireland, Italy and Portugal (see Gabriel, 1992, p. 462f).

2. In all, 2,960 citizens were asked to complete the questionnaire. The main subject of the survey was cultural policies, not sports policies. The representativenesss of the sample was checked against age, sex, religion and education (Gabriel et al. 1993b, page 6ff). For these indicators, the 476 respondents reflected the city's residents on the first three, but over-represented those with more formal education. Analysis of responses showed no difference in spending preference by education for four of the nine sports policies investigated here. For the other five, there was a slight tendency for those at the lower end of the eduation spectrum to favor higher levels of spending. In interpreting the results given below, it is worth keeping in mind that, on this account, the citizen sample might slightly under-represent the distance between citizen and councilor spending preferences on sports policies.

3. For different sport policies, the improtance of an issue was determined from citizen responses to the question: "How important are the following local sport policies in your opinion?" The heterogeneity of views was determiend by the standard deviation of the public assessments for spending preferneces on each of the nine sports activities.

4. Here, proximity $= \dfrac{\Sigma \text{ (position of the ith constituent} - \text{position of the local council})^2}{\text{number of constituents}}$

5. Centerism $=$ proximity $- \dfrac{\Sigma \text{ (position of the ith constituent} - \text{public opinion})^2}{\text{number of constituents} - 1}$

REFERENCES

Achen, C. (1977). Measuring Representation: Perils of the Correlation Coefficient. *American Journal of Political Science 21*, 805–815.
Achen, C. (1978). Measuring Representation. *American Journal of Political Science 22*, 475–510.
Arzberger, K. (1980). Bürger und Eliten in der Kommunalpolitik. Stuttgart: Kohlhammer.
Backstrom, C. H. (1977). Congress and the Public: How Representative Is the One of the Other? American *Politics Quarterly 5*, 411–435.
Blalock, H. M. (1967). Causal Inferences, Closed Populations and Measures of Association. *American Political Science Review 61*, 130–136.
Brettschneider, F. (1995). *Öffentliche Meinung und Politik – Eine empirische Studie zur Responsivität des Deutschen Bundestages zwischen 1949–1991*. Opladen: Westdeutscher Verlag.
Brettschneider, F. (1996). Responsiveness of the German Bundestag in Comparative Perspective. Paper for the Annual Meeting of the World Association for Public Opinion Research. Salt Lake City, Utah.
Brettschneider, F., & Walter, M. (1996). Ratsmitglieder und Bürger. Paper for the Conference Kommunalpolitik und Lokale Politikforschung – Alte Probleme und Neue Herausforderungen an der Schwelle des Jahres 2000, Heidelberg.
Brooks, J. E. (1985). Democratic Frustration in the Anglo-American Polities: A Quantification of Inconsistency between Mass Public Opinion and Public Policy. *Western Political Quarterly 38*, 250–261.
Brooks, J. E. (1987). The Opinion-Policy Nexus in France: Do Institutions and Ideology make a Difference? *Journal of Politics 49*, 465–480.
Brooks, J. E. (1990).The Opinion-Policy Nexus in Germany. *Public Opinion Quarterly 54*, 508–529.

Clausen, A. R. (1977). The Accuracy of Leader Perceptions of Constituency Views. *Legislative Studies Quarterly 2*, 361–384.

Cnudde, C. F., & McCrone, D. J. (1966). The Linkage Between Constituency Attitudes and Congressional Voting Behavior: A Causal Model. *American Political Science Review 60*, 66–72.

Cnudde, C. F., & Neubauer, D. E. (1969). New Trends in Democratic Theory. In: C. F. Cnudde, & D. E. Neubauer (Eds.), *Empirical Democratic Theory*, (511–534). Chicago: Markham.

Dahl, R. A. (1971). *Polyarchy: Participation and Opposition*. New Haven: Yale University Press.

Dalton, R. J. (1985). Political Parties and Political Representation. Party Supporters and Party Elites in Nine Nations. *Comparative Political Studies 18*, 267–299.

Downs, A. (1957). *An Economic Theory of Democracy*, New York: Harper and Row.

Eulau, H., Wahlke, J. C., Buchanan, W., & Ferguson, L .C. (1959). The Role of the Representative: Some Empirical Observations on the Theory of Edmund Burke. *American Political Science Review 53*, 742–756.

Gabriel, O. W. (1992). Wertwandel, Kommunale Lebensbedingungen und die Aufgaben der Kommunen am Beginn der Neunziger Jahre – Eine Empirische Analyse Politischer Einstellungen Kommunaler Mandatsträger in 14 Städten. In: F. Schuster, & G. W. Dill (Eds.), *Kommunale Aufgaben im Wandel*, (149–235). Köln: Kohlhammer.

Gabriel, O. W., Brettschneider, F., & Kunz, V. (1993a). 'Responsivität Bundesdeutscher Kommunalpolitiker.' *Politische Vierteljahresschrift 34*, 29–46.

Gabriel, O. W., Brettschneider, F., & Walter, M. (1993b). Kultur in Stuttgart – Ergebnisse einer Bevölkerungsumfrage im Herbst 1993. Paper available from the Department of Political Science, University of Stuttgart.

Hoffmann-Lange, U. (1991). Kongruenzen in den Politischen Einstellungen von Eliten und Bevölkerung als Indikator für Politische Repräsentation. In: H. D. Klingemann, R. Stöss, & B. Wessels (Eds.), *Politische Klasse und Poltitische Institutionen. Probleme und Perspektiven der Elitenforschung*, (275–289). Opladen: Westdeutscher Verlag.

Jackson, R. A. (1992). Effects of Public Opinion and Political System Characteristics on State Policy Outputs. *Journal of Federalism 22*, 31–46.

Kessel, J. H. (1964). The Washington Congressional Delegation. *Midwest Journal of Political Science 8*, 1–21.

Kingdon, J. W. (1973). *Congressmen's Voting Decisions*. New York: Harper and Row.

Kurer, P. (1979). *Repräsentation im Gesetzgebungsverfahren: Der Einfluß der Amerikanischen Wähler auf den Kongreß*. Zürich: Schultheiss Polygraphischer Verlag.

Lijphart, A. (1984). *Democracies. Patterns of Majoritarian and Consensus Government in Twenty-One Countries*. New Haven: Yale University Press.

McCrone, D. J., & Stone, W. J. (1986). The Structure of Constituency Representation: On Theory and Method. *Journal of Politics 48*, 956–975.

Miller, W. E., & Stokes, D. E. (1963). Constituency Influence in Congress. *American Political Science Review 57*, 45–56.

Monroe, A. D. (1979). Consistency Between Public Preferences and National Political Decisions. *American Politics Quarterly 7*, 3–19.

Pennock, J. R. (1952). Responsiveness, Responsibility and Majority Rule. *American Political Science Review 46*, 790–807.

Pitkin, H. F. (1967). *The Concept of Representation*. Berkeley: University of California Press.

Powell, G. B., & Powell, L. W. (1978). The Analysis of Citizen-Elite Linkages: Representation by Austrian Local Elites. In: S. Verba & L.W. Pye (Eds.),*The Citizen and Politics: A Comparative Perspective*, (197–217). Stanford, Connecticut: Greylock.

Sharpe, L. J. (1970). Theories and Values of Local Government. *Political Studies 18*, 153–174.
Weissberg, R. (1978). Collective vs. Dyadic Representation in Congress. *American Political Science Review 72*, 535–547.
Wieland, H., & Rütten, A. (1991). *Sport und Freizeit in Stuttgart – Sozialempirische Erhebung zur Sportnachfrage in Einer Großstadt.* Stuttgart: Naglschmid.

7. PLEDGES AND PERFORMANCE: AN EMPIRICAL ANALYSIS OF THE RELATIONSHIP BETWEEN MANIFESTO COMMITMENTS AND LOCAL SERVICES

Rachel E. Ashworth and George A. Boyne

Mandate theory suggests that political parties translate their ideological beliefs into specific pledges that are incorporated within party manifestos and form the basis of election campaigns. Once elected, it is further assumed that these policy pledges are enacted by political parties in government. This theory has frequently been operationalized at national government level, in the U.K. and elsewhere by examining the extent of specific pledge implementation (Pomper & Lederman, 1980; Rose, 1980; Rallings, 1987; Royed, 1996). More recently, a revised test, based on the relationship between pre-election policy emphases and governmental expenditure shares has been applied (Petry, 1988; Hofferbert & Klingemann, 1990; Budge & Hofferbert, 1990, Hofferbert & Budge, 1992). However, there has been no analysis of the relationship between manifesto pledges and subsequent service performance, either at national or local levels. In this chapter we begin to fill this gap in the literature by assessing the relationship between election pledges and service performance in Welsh local government.

MANDATE THEORY

Democratic theory suggests that differing views within society can be expressed through the political system. During election campaigns, political parties representing these differences appeal to groups of the electorate for votes by offering a set of policies they hope will put them in office. Thus, when a party has enough votes to take office it is said to have a 'mandate' to implement its electoral program. In theory, the electorate faces a choice between parties and their specific policies, and gives the winning party a mandate to govern, based upon the policy content of its manifesto. It has been suggested that the health of a democracy can be assessed by testing the relationship between manifesto content and government policy. As one example, Strom & Leipart (1989, p. 265) describe the manifesto as the ". . . key link in the democratic interaction between the governors and the governed."

Yet there is debate over the role of the manifesto and consequently the credibility of mandate theory. Critics suggest that manifestos are nothing more than 'wish lists' issued by unrealistic opposition parties who, nevertheless, will be committed to fulfilling this program once elected, however inappropriate it may be. Finer (1975, p. 379) suggests that the U.K., in particular, suffers from 'manifesto-itis' and is critical of parties blindly implementing pledges:

> The mystique which a grotesque perversion of democratic theory has imparted to these documents, together with the multifarious details of the commitments they nowadays contain, conduces to bad government and the discredit of the parliamentary system as a whole.

The credibility of mandate theory is further weakened by the implausibility of the assumption that the electorate votes on the basis of the content of manifestos. In recent years there has been considerable growth in both the length and specificity of party manifestos in the U.K., both nationally and locally (e.g. a 70,000 word Labour Manifesto was produced for the GLC elections in 1980). It is unlikely that, when voting for a political party, electors concur with all commitments in one manifesto and disagree with all pledges made by other parties. Indeed, the majority of electors do not even read a party manifesto. However, most electors have access to a mini-manifesto from their local candidate. Also, voters are subjected to media discussion throughout election campaigns that usually center on policy pledges (Robertson, 1976; Topf, 1994). Thus, whilst there remains some debate concerning the electorate's understanding of the policy content of manifestos, many tests of mandate theory have continued to be based on the belief that, whether read or not, manifestos provide a clear indication of a party's policy intentions. As Robertson (1976, p. 72) states:

... party manifestos ... are the only direct and clear statements of party policy available to the electorate and directly attributed to the party as such. Though it is perhaps unlikely that many voters read them themselves, they are the source and official backing for any impression that the electorate may get of what the parties stand for ... they are discussed and represented by the speeches of party leaders and they are the basic source for the campaigns of constituency candidates.

PRIOR WORK

Previous research on mandate theory has almost entirely focused upon national governments, with the exception of a state-level analysis in the USA (Elling, 1979) and an examination of the implementation of several policies contained in the 1978 Labour Manifesto for the London Borough of Brent (Fudge, 1981). Prior studies can be grouped into two categories. The first consists of tests of the implementation of specific manifesto commitments, involving a comparison between specific election promises and government policy actions (Ginsberg, 1976; Elling, 1979; Pomper & Lederman, 1980; Rose, 1980; Rallings, 1987; Kalogeropoulou, 1989; Royed, 1996). 'Specific' election pledges consist of firm commitments to policies, such as "We will increase expenditure on education," whilst vague statements on policy, such as "We believe in nursery education for all," are broadly recognized as 'rhetorical' pledges. The second category of work focuses on the measurement of the relationship between pre-election policy emphases and post-election budgetary allocations (Petry, 1988; Budge & Hofferbert, 1990; Hofferbert & Klingemann, 1990; Hofferbert & Budge, 1992). The overwhelming majority of empirical evidence produced by both types of test upholds the validity of mandate theory at the national government level. However, it is important to consider weaknesses associated with each test that may have influenced these results.

Tests Based Upon the Extent of Specific Pledge Implementation

In this approach the proportion of a manifesto that contains rhetorical and vague statements is disregarded. As specific pledges tend to focus upon mainly peripheral policy areas, this tends to mean that statements on major policy areas are discarded. Consequently the analysis of "pledges may not provide a complete basis for judging party or government compliance with election promises." (Hofferbert & Budge, 1990, p. 112). However, Royed (1996, p. 54) argues that "... the best way to 'test' the mandate theory of elections is to look only at firm pledges, and to determine whether or not each pledge has indeed been redeemed." Similarly, Kalogeropoulou (1989) maintains that his work on the implementation of the 1981 PASOK manifesto had to be based upon specific pledges as PASOK had made a firm commitment to reform major Greek policies and institutions. Ginsberg (1976, p. 45) agrees:

> The analysis of statutes rather than some other form of policy output (governmental expenditures, for example) appeared to represent the method best suited to the purpose of this study. Since the aim was to determine whether the pre-election positions of policy-makers were implemented in public policies, it appeared to us to be necessary to look directly at the substance of the relevant public policies.

There is some debate over whether indicators such as government expenditure shares accurately represent government policy (see below). There is less uncertainty regarding the interpretation of specific action as government policy. Moreover, Hogwood & Gunn (1984, p. 15) suggest that "... statements of specific actions which political organizations (interest groups, parties, the Cabinet itself) would like to see undertaken by government" indicate policy. To emphasize their case they cite examples from the 1979 Conservative Party Manifesto: "We shall give union members the right to hold ballots for the election of governing bodies of trade unions" (p. 15). They suggest that such statements may be part of a broader policy picture but that they can stand as policy proposals in their own right.

Previous studies looked to a variety of data sources for evidence of pledge implementation. Many focused purely on legislation (Ginsberg, 1976; Elling, 1979; Rose, 1980), whilst others incorporated legislation as just one of a number of indicators, including the Queen's Speech, budgetary and economic statistics, executive actions and party literature (Rallings, 1987; Kalogeropoulou, 1989; Royed, 1996). Clearly, the limitation of legislation as an indicator of policy is that it can only represent the first stage of a policy action. As Rose (1980, p. 67) suggests, legislation is "... a means to an end. But there is no assurance that the chosen government's means will be sufficient to achieve the desired social ends." Rallings (1987) cites the example of the Equal Pay Act of 1970 as an example of fulfilment of a pledge but a failure of policy, with large sectors of industry ignoring the provisions of the Act.

Tests Based Upon the Relationship Between Pre-Election Policy Emphases and Governmental Expenditure Shares

Problems with this test concern the presumed relationship between pre-election policy priorities and expenditure shares. Royed (1996) questions the validity of the assumption that the number of sentences devoted to a particular policy area in an election program should be associated with the proportionate expenditure allocation for that policy area. This is a far from straightforward assumption as, whilst the number of sentences devoted to each policy area provides a good indication of a party's policy priorities, pledges may denote expenditure reductions. The fear is that some pledges could be categorized contrary to their meaning. Royed emphasizes that the calculation of policy emphases could include pledges that increase, decrease or maintain existing levels of expenditure. This, he suggests, indicates "... positive, negative or no correlation between policy emphases and spending could all be quite consistent with mandate theory" (Royed, 1996, p. 53). Set against this, Hofferbert & Budge (1992) maintain that they have rarely found any sentences pledging cuts in services or expenditure, as most parties simply de-emphasized topics rather than make negative pledges. Despite using both 'pro' and 'anti' categories in previous work (Budge et al. 1987), they found that the overwhelming proportion of references were in the 'pro' category. Thus, they maintain that if there is a heavy emphasis on, for example, social services in a manifesto, there is a logical assumption that the party will extend social services once in government.

There are further doubts about this test of mandate theory. For one, the Budge & Hofferbert method gives equal weight to rhetorical and specific pledges. Whilst one can accept the premise that rhetoric could be positively associated with expenditure allocations, Royed (1996, p. 52) remains concerned with the fulfilment of specific pledges: "... this is important for mandate theory." He maintains that an association between party emphases and governmental expenditure is not sufficient support for mandate theory. He suggests that the interpretation of such results as evidence of mandate theory is misleading on the grounds that, despite a relationship between policy emphases and expenditure, specific pledges could remain unfulfilled. This makes it difficult to state categorically that manifestos impact on policy. Royed (1996, p. 54) concludes that this operationalization of mandate theory is "... at best only a very rough estimate of the relationship between party programme commitments and policy action." Yet Petry (1988, p. 386) maintains that "... expenditure data are a better source of evidence on government decision priorities than legislation'. Nevertheless, he suggests that his findings – a rejection of mandate theory on the basis of a lack of association between party emphases and governmental expenditure in post-war Canada – could differ significantly with the inclusion of non-budgetary pledges.

There are further objections related to the use of expenditure levels or shares as a measure for government policy. Hofferbert & Klingemann (1990, p. 286) readily admit that "... expenditures are not the perfect indicators for a broader concept of government behaviour or policy" and accept that their work excludes legislation and regulations which are not covered by expenditure data. Other authors are equally critical of the use of expenditure as an indication of government policy. Royed suggests that aggregate-spending totals could disguise small changes resulting from the implementation of specific pledges and provides the example of an alteration in eligibility criteria for a benefit. Similarly, Ginsberg (1976, p. 45) argues that:

> Governmental expenditures offer an extremely reasonable type of indicator and are extremely useful for some purposes. But, expenditures are based, at least in part, on the actual cost of effecting policy aims. Differential costs of implementing various aims make comparisons across types of policy difficult.

Hofferbert & Budge (1992) accept that spending on an area such as regulation may not provide an indication of its relative importance compared with social security or health, so can an association between policy emphases and budget shares indicate the true impact of manifesto pledges on policy? It is more likely that the relationship between the two could indicate merely the first stage of a commitment to a policy area. This is consistent with Hofferbert & Klingemann's (1990, p. 286) statement on the use of expenditure shares as policy indicators: "Our analysis is but one aspect of the process, but we would argue that it is a consequential aspect." Overall, it would seem that expenditure data can at best provide an indication of the direction of government policy and, as Hofferbert & Budge (1990) suggest, is just one of a key set of linkages between party emphases and a policy product.

Thus, prior studies have operationalized mandate theory either by checking on the implementation of specific pledges or by measuring the relationship between pre-election policy emphases and governmental expenditure priorities. There are weaknesses associated with both tests, but perhaps the most important is the failure to assess the link between manifesto pledges and the substantive achievements of government, as indicated by the performance of public services.

Mandate Theory at the Local Level

There has been very little analysis of local manifesto content, let alone any assessment of its impact on policy-making (Ashworth, 2000b). Most work has focused on the formulation and content of policy documents (Game & Skelcher, 1983; Gyford et al. 1989). Among the studies that do exist at the local level, Wilson & Game (1998) examined manifestos produced for three local elections (Richmond-upon-Thames and Wandsworth in 1990, and Edinburgh in 1992). They concluded that political parties and their election programs do help determine public policy even at a local level:

> All the significant candidates stood under their respective party labels, embraced publicly their parties' manifesto and campaign promises, which in turn determined the key topics of electoral debate. The parties, in short, defined and focused the campaign and probably

boosted whatever interest the media and electorate might have had in it. (Wilson & Game 1998, p. 275)

There has been little investigation of mandate theory in U.K. local government. However, there is evidence to suggest that for some Labour and Liberal Democrat authorities the manifesto has been officially adopted as council policy, with some councils measuring progress via manifesto checklists or monitoring groups (Gyford et al. 1989). This is supported by Stoker (1991), who cites the results of the Widdicombe survey of 1986, which indicated that 63% of councilors believe that their main priority is the fulfilment of manifesto pledges. The figure rises to 87% for Labour councilors.

An analysis of the fulfilment of local manifesto pledges in one U.K. local authority is provided by Fudge (1981), who adopted a case-study approach in order to examine the formulation, content and implementation of the 1978 Labour Manifesto for the London Borough of Brent. His 'partial' account draws on information obtained from 'haphazard' semi-structured interviews, documentation, meetings and 'political gossip', aimed at determining whether local policy-making reflected a pre-determined party agenda or a response to local issues. He concluded that whilst the manifesto lacked overall policy priorities and a costing of promised programs, most party members felt that objectives had been achieved.

A more recent study on Welsh local elections involved the application of both conventional tests of mandate theory (Ashworth, 2000a). The first involved measuring the extent to which specific manifesto commitments were fulfilled. The empirical evidence suggests that mandate theory is upheld as a slight majority of pledges (52%) were fulfilled in full or part, with future implementation planned for a further 3%. Certain service areas demonstrated high fulfilment rates – Education with 60%, Social Services with 62% and Economic Development with 72%. Others were characterized by low levels of implementation – Local Government Reorganization and Local Democracy with 37%, Highways with 22% and Housing with 20%. Further analysis of sub-service pledge fulfilment revealed high levels of implementation in certain policy areas, including Children's Services, Special Needs, Police and Crime, Waste Collection, Skills Development, Job Creation and Library Services. Across services there was a high degree of fulfilment in relation to strategy-based pledges, particularly in relation to Environmental and Planning Services, Education, Leisure Services or Economic Development. At the individual local authority level, the majority of political parties have implemented over half of their specific election promises, either in full or part. Parties in some authorities have fulfilled a substantial proportion of pledges – for example, Flintshire with 59%, Carmarthenshire with 62% and Gwynedd with 70%.

The second test, based on the relationship between pre-election policy emphases and expenditure priorities proved inconclusive. There is evidence of a positive association between expenditure and party pledges at an aggregate level, especially between Labour pledges and total expenditure. However, evidence based on the relationship between service-level expenditure and equivalent party pledges offers little support for mandate theory. Despite this, results on the relationship between policy emphases and expenditure priorities in single authorities provides some support for mandate theory. For example, the Labour manifesto for the Vale of Glamorgan local elections in 1995 is positively related to the authority's subsequent expenditure priorities, according to results based on Pearson and Spearman coefficients. However, for the majority of authorities, whilst there is a positive relationship, it is not significant. Whilst a positive relationship exists between pledges and expenditure at the aggregate level, this was not the case for service-level expenditure. Neither was there a consistently positive and significant relationship between policy emphases and expenditure allocations at the local authority level. Thus, the empirical evidence from the operationalization of mandate theory based on Welsh local elections proves conflicting, with any significant positive relationships between manifesto pledges and expenditure evident at the aggregate level only.

In sum, previous studies have operationalized mandate theory either by checking on the implementation of specific pledges or by measuring the relationship between pre-election policy emphases and governmental expenditure priorities. However, there are many definitions of policy and many more definitions of performance. Local authorities, along with most public sector organizations in the U.K., are frequently subjected to performance measurement, whether by central government, the Audit Commission or from within. Such performance measures are continually being improved with the focus now firmly on policy outputs and, to a lesser extent, policy outcomes. Policy outputs can include "... the payment of cash benefits, the delivery of goods and services, the enforcement of rules, the invocation of symbols or the collection of taxes" (Hogwood & Gunn, 1984, p. 16). Performance indicators that are based on proportions of funds/staff or on physical resources fall into the Hogwood and Gunn category of 'intermediate output'. The impact of policy emphases on service performance, therefore, constitutes an alternative but as yet untried operationalization of mandate theory.

METHODOLOGY

The aim of this research is to evaluate the relationship between election promises and local government service performance. The methodology comprises multivariate statistical analysis with local authority performance as the dependent variable. Manifesto pledges are the key explanatory variable, although variables representing total expenditure and party control will also be included within the multivariate analysis. In this preliminary analysis we focus on the 22 unitary authorities in Wales. This has the advantage of covering all local government services, and effectively holds constant potential regional effects on performance (e.g. political culture, institutional arrangements).

Measurement of Variables

Manifesto Pledges

Manifesto pledges were obtained from documents produced for the 1995 Welsh local elections. These were the first elections for the 22 new unitary councils that came into operation in April 1996. These replaced the former eight counties and 37 districts with a single tier of local government. The unitary councils are elected in full every four years, rather than in thirds, as is the case with some English councils. Although turnout for local elections in the UK is traditionally low, there was a higher than average turnout of almost 50% for the Welsh unitary elections in 1995. It has been argued that this set of elections acted as a stimulus to party competition in Welsh local government (Rallings & Thrasher, 1997).

The manifesto documents were subject to a content analysis which is "... a research technique for the objective, systematic and quantitative description of the manifest content of communication" (Berelson, 1952, p. 18). As with all prior tests (except Ginsberg, 1976), the recording unit consisted of a sentence or quasi-sentence, with each coded as a possible pledge. In relation to pledge coding and categorization, content analysis demands the replication or modification of prior schemes wherever possible. Thus, previous work guided the categorization of 'rhetorical' and 'specific' pledges (Rallings, 1987; Royed, 1996). Vague statements in support of a policy or service area were coded as 'rhetorical pledges'; for example, 'We wish to protect the environment in which we live and work'. More detailed policy commitments were coded as 'specific pledges' – these included statements such as 'We will establish an Environmental Protection Committee'. These were further categorized according to the nature of the action pledged; for example, whether the pledge stipulated 'direct action', 'increases in expenditure', etc.

There were difficulties in relation to policy categorization, as prior analyses of manifesto content had been based on national government activities, with pledges coded according to broad policy domains facilitating longitudinal or cross-country comparisons. For this research, alternative policy categories were devised along the lines of local government services such as education, housing and social services. Further categories were developed within services; for example, in Education, the sub-service categories included nursery, primary, secondary and special needs. In an attempt to collect the entire 'universe' of data (i.e. all manifestos produced for the 1995 elections), 55 documents were analyzed, as produced by the four main political parties in Wales and by Independent candidates. Manifestos were unavailable for just one authority, namely Conwy.

Two reliability tests were applied to a sample of manifesto content, constituting around 8% of the text under analysis. The first test was based on stability, ensuring that the coding scheme remained 'invariant'. The test involved the researcher coding and re-coding the data on the same sample of text until coding stability was achieved. Whilst there is no stipulated reliability score that must be achieved, most researchers are satisfied with a reliability score of 90% or above, although clearly 100% stability would be ideal (Weber, 1990). The evidence from the stability tests revealed reliability scores of 94% and 93% for 'action pledged' and 'pledge categorization', respectively. These results can be compared to those from Petry (1988), who reported a stability rate of 90%. Whilst stability tests are useful indicators of reliability, they cannot be solely relied upon, particularly in cases where one person is coding the data (Krippendorff, 1980; Weber, 1990). Thus, a second test was applied, based on 'reproducibility'. This involves coding of the same sample of text by two different coders, until agreement is reached. Again, the results indicate reliability scores of over 90%, with 99% and 93% for 'action pledged' and 'pledge categorization', respectively. This evidence compares favorably with reported results from prior tests. Thus, Ginsberg (1976) achieved reproducibility results of 0.94, although the test was based on just one manifesto, whilst Royed (1996) reported scores of 96% for 'policy categorization' and just 85% for 'type of pledge'.

Ideally, 100% reliability scores should be achieved, as only then would the conditions for reliability be entirely satisfied. However, with one reproducibility result of 99% and the remainder over 90%, the scores appear to satisfy reliability criteria. On this basis, and following the general recommendation from Krippendorff (1980) and Weber (1990), that content analysis procedures should be at least stable and reproducible, the reliability scores uphold the internal validity of the coding and categorization procedure for the Welsh local manifestos. In relation to external validity, the content analysis satisfies the criteria for hypothesis or construct validity which, according to Krippendorff (1980, p. 158):

> ... assesses the degree to which an analytical procedure models, mimics or functionally represents relations in the context of data. In content analysis this form of validity is

concerned principally with the nature of the analytical construct which is accepted or rejected on the basis of a demonstrated structural-functional correspondence of the processes and categories of an analysis with accepted theories, models and knowledge of the context from which data stem.

For the purposes of this study, it is only winning party pledges that are included in the analysis, since mandate theory is primarily concerned with the impact of the winning manifesto. Also, in order to represent accurately the entire content of the manifesto, rhetorical and specific pledges are combined to provide a total number of pledges. Finally, the focus is on pledges at service level, such as the proportion of pledges on education. On the basis of mandate theory and prior evidence of links between pledges and policies, we hypothesize a positive relationship between the proportion of pledges on a service and subsequent performance.

Table 7.1 contains descriptive statistics on proportions of election pledges devoted to particular service areas. The greatest proportion of pledges focus on Environment and Planning services, with many specific commitments, such as "We will publish an Environmental Charter for the area to indicate commitments to environmental improvements and monitoring" (Plaid Cymru – Gwynedd). On average, around 17% of pledges were made on Education, ranging from vague statements, such as "Labour is committed to providing a quality education for all" (Labour – Torfaen), to more specific commitments, like: "All rising four year olds whose parents desire it can have a nursery place in Flintshire" (Labour – Flintshire). A high proportion of pledges (15%) centered on Leisure, mainly demonstrating a commitment to library services: "A Labour Carmarthenshire will be committed to improving the use of libraries with longer opening hours" (Labour – Carmarthenshire). Social Service pledges constituted around 12% of the total, the majority of which concerned community care issues: "Labour is committed to a flexible home care service providing a range of care in the early morning and evening with emergency cover during the night" (Labour – Swansea). Highways pledges, in general, centered on road maintenance and safety issues but did include some more ambitious policy objectives: "The new council will seek funds for the construction and stocking of at least two cross-city public transport links, either light rail, guided bus or tram" (Labour – Cardiff). Just over 10% of pledges related to Housing, mostly housing maintenance: "Massive investment will continue, repairing and maintaining properties, as well as a programme for the installation of UPVC windows" (Labour – Torfaen), as well as tenant rights: "Efforts should be made towards the creation of a Federation of Tenants Associations which could represent the views of tenants across the new County Borough on major policy matters" (Labour – Vale of Glamorgan). A smaller proportion of pledges (7%)

Table 7.1 Descriptive Statistics – Proportions of Election Pledges

Service	Mean	C.V.[1]	Minimum	Maximum	n
Education	16.92	54.43	3.21 (Cardiff)	34.14 (Vale)	20
Social Services	11.99	67.30	3.21 (Cardiff)	26.00 (Merthyr)	18
Housing	10.59	83.76	1.79 (Denbighshire)	28.57 (B. Gwent)	18
Highways	11.50	70.69	4.44 (Anglesey)	26.61 (Cardiff)	18
Leisure	15.55	148.42	1.81 (Vale)	50.00 (Bridgend)	18
Environmental	18.36	77.99	4.55 (Bridgend)	69.23 (Ceredigion)	20
Protective	7.00	129.57	2.44 (Flintshire)	35.71 (B. Gwent)	16
Finance	6.10	151.31	9.23 (Gwynedd)	40.91 (Bridgend)	15

C.V. = Coefficient of Variation.

was devoted to Protective Services, with this pledge from the Labour Party typical of many: "We will seek to educate consumers and traders about their statutory rights and obligations through a series of exhibitions" (Labour – Vale of Glamorgan). Finally, just 6% of pledges focused on Finance. These varied from those aiming for greater local efficiency, to commitments to reform local financial systems: "Plaid Cymru are committed to ensuring a fairer rating system for business" (Plaid Cymru – Gwynedd).

Service Performance

Despite the proliferation of performance indicators in the public sector over the last decade, there has been little theoretical reflection on the concept of organizational performance. Much of the literature on performance indicators consists of criticisms of their purpose and content (e.g. Cutler & Waine, 1997). The underlying interpretations of performance are implicit or underdeveloped (a rare exception is Carter et al., 1992).

An earlier body of research, however, provides a useful starting point for a discussion of the meaning of organizational performance. This is the literature on organizational 'effectiveness' that emerged during the 1960s and 1970s, largely in sociology and management journals in the USA. The initial empirical work in this area was based on narrow definitions of performance, and used only a handful of measures of the concept (Steers, 1975). However, it was soon recognized that organizational performance is multi-dimensional, and that many measures are required in order to reflect this complexity. In particular, performance is likely to be viewed differently by different stakeholders (Kimberley et al., 1983; Cameron, 1986). It is, in short, politically constructed. This implies that no technically correct interpretation of the concept exists: good (or bad) performance is in the eye of the beholder. Thus high-performing organizations can be described as those that satisfy the demands of their key stakeholders (Connolly et al. 1980); an insight that has been rediscovered by writers on public administration in recent years (e.g. Ott & Shafritz, 1994; Stewart & Walsh, 1994).

This political perspective on the performance of organizations is encapsulated in the 'competing values' model developed by Quinn & Rohrbaugh (1981, 1983). They argue that the main features of performance can be arrayed on two axes. First, the locus of power to define organizational performance: does this rest predominantly inside or outside the organization? Secondly, the attitudes of key stakeholders towards control and flexibility: is there a need to prescribe processes and performance targets, or should managers be given autonomy and 'left to get on with it'?

These two axes are combined in Figure 7.1, which contains four 'ideal type' models of performance assessment. Rohrbaugh (1983) argues that these models represent distinctive research traditions in the social sciences. Equally, they represent different approaches to the external and internal management of public agencies. The 'human relations' model combines an internal locus of power with an emphasis on organizational flexibility. The major performance criteria here refer to the cohesion, morale and development of the workforce. The 'open

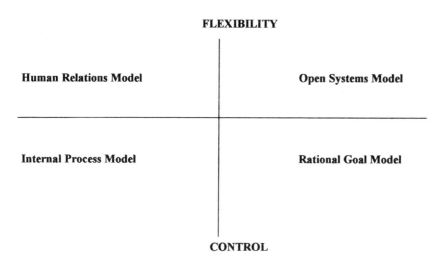

Figure 7.1: Models of Organizational Performance
Source: Rohrbaugh (1983)

system' model also emphasizes organizational autonomy over means, but in the context of ends that are externally set. Rohrbaugh (1983) argues that relevant performance criteria from this perspective include growth, resource acquisition and political support. The 'internal process' model combines autonomy from external constraints with an emphasis on control within the organization. Important performance criteria include 'an orderly work situation' and the provision of a 'sense of continuity and security' for members of the organization (Rohrbaugh, 1983, p. 269). Finally the 'rational goal' model comprises an external locus of power, tight regulations on organizational goals, and performance criteria such as productivity and efficiency.

In any organization, all of these models may be present to some extent. Nevertheless, some political groups, and the performance criteria that they prefer, are likely to be predominant. In recent years public organizations in the U.K. and elsewhere have been more heavily regulated by governmental agencies (Hood et al., 1998), thereby shifting the locus of power to define performance towards the external rather the internal end of the horizontal axis in Figure 7.1. Furthermore, performance indicators and targets have been increasingly applied to whole sets of organizations, such as schools and hospitals, so there has been a shift down the vertical axis towards 'results-oriented' management (Durant, 1999). Together these movements constitute an emphasis on the achievement of explicit and quantified standards of service provision that have been identified by a powerful external stakeholder, that is, central government.

The rational goal model is reflected strongly in the performance measurement regime that has been established for local governemnt in England and Wales. Under the 1992 Local Government Act, the Audit Commission (a body established by central government) has powers to specify a set of performance indicators (PIs) that local authorities must collect and publish. The indicators from this data set that correspond most closely to the concept of organization performance are listed in Table 7.2 and in Table 7.3 for financial years 1996/97 and 1997/98 respectively. The data covers the first two years of the new unitary councils (see Boyne, 1997, for a comprehensive analysis of the Audit Commission PIs). Although structural reorganization may offer local policy makers an ideal opportunity to 'start afresh', levels of performance in the initial period of the unitary authorities may have been largely inherited from their predecessors, and therefore show little relationship with manifesto commitments. In order to circumvent this problem, we have also measured short-term shifts in performance from 1996/97 to 1997/98 (see Table 7.4).

These indicators include all aspects of the major local government services (e.g. education, social services, housing). They also relate to important dimensions of performance such as service *coverage* (e.g. number of physically disabled people receiving care), service *quality* (e.g. number of homeless households in temporary accomodation), service *speed* (time taken to issue special educational need statements), service *efficiency* (e.g. cost per household of refuse collection), service *utilization* (e.g. visits to libraries) and administrative *effectiveness* (e.g. percentage of potential council tax revenue that is actually collected).

Both levels and changes in service performance show substantial variations for all 26 of the PIs that we analyze. For example, the percentage of student grants paid on time varies from 29% to 100% (see indicator E6 on Table 7.2), and the cost per household of refuse collection varies from £14 to £53 (see indicator EN3 in Table 7.2). Similarly, there are cases of increases and decreases in service standards for each PI between 1996/97 and 1997/98. For example, the average length of stay in temporary accommodation by homeless households rose by 10 weeks in Bridgend but fell by the same amount in Denbighshire (see indicator H1 in Table 7.4).

Thus, local service performance appears to vary widely, and it is possible that different manifesto commitments by different councils can account partly for this. However, manifesto pledges cannot provide an equally good explanation of all these variations in performance, simply because the PIs themselves are not closely related to each other. The correlation results in Tables 7.5 to 7.18 show very few significant relationships between PIs in the same service area for either 1996/97 or 1997/98. Furthermore, the coefficients are as likely to be negative as positive, which suggests that above average performance in one aspect of service provision is likely to be accompanied by below average performance in another. Nor are the PIs moving uniformly up or down over time (see, for example, Tables 7.7, 7.11, 7.13 and 7.16). This pattern of evidence shows that no single variable can provide a comprehensive explanation of differences in performance.

Control Variables

Although mandate theory suggests that party pledges are a significant influence on service performance, they are unlikely to be the only or even a dominant influence. If is therefore important to take other explanatory variables into account. There is, however, no comprehensive model of public service performance from which such variables can be derived. In this explanatory analysis we incorporate two variables that have played a prominent role in studies of local policy variation: the level of financial resources and the party political composition of the local council.

Table 7.2 Descriptive Statistics: Performance Indicators 1996/97

Service	Mean	C.V.	Minimum	Maximum	n
Education					
E1. % Children under five in LA schools	79.11	17.16	56 (Powys)	100 (Wrexham)	22
E2. % of primary pupils admitted in excess of school capacity	5.31	75.32	1 (Anglesey)	16 (B. Gwent)	22
E3. % of secondary pupils admitted in excess of school capacity	1.79	153.63	0 (Anglesey)	9 (Newport)	22
E4. % of Special Need Statements issued within 18 weeks	43.48	61.79	1 (Carmarthenshire)	86 (Merthyr Tydfil)	21
E5. Enrolments in adult education per 1000 pop	23.75	85.43	0 (Flintshire)	55 (Bridgend)	20
E6. % Mandatory student awards paid on time	91.57	18.69	29 (Conwy)	100 (Pembs)	21
Social Services					
SS1. Number of physically disabled receiving care, per 1000 adults	5.24	152.29	0.38 (Neath PT)	36.48 (Merthyr)	21
SS2. Number of people with learning difficulties receiving care, per 1000 adults	3.82	80.37	0.26 (Conwy)	15.70 (B. Gwent)	22
SS3. Number of people with mental health problems receiving care, per 1000 adults	4.35	127.12	0.11 (Flintshire)	21.12 (Anglesey)	21
Housing					
H1. Average number of households in temporary accommodation, per 1000 households	3.42	124.85	0.60 (Bridgend)	24 (Flintshire)	22
H2. Average length of stay in temporary accommodation (weeks)	7.10	69.29	1 (Torfaen)	23 (Monmouth)	21

Table 7.2 Continued

Service	Mean	C.V.	Minimum	Maximum	n
H3. % of Housing Benefit claims processed in 14 days	81.67	21.84	25 (Denbighshire)	99 (B. Gwent)	21
H4. Cost of Housing Benefit administration per claimant	51.75	40.01	18 (Conwy)	87 (Cardiff)	21
H5. % of tenants owing 13 weeks+ rent	3.08	87.98	1 (Monmouth)	10 (Denbighshire)	22
H6. Average time taken to relet/refurbish dwellings	5.04	43.45	2 (Powys)	11 (Torfaen)	21
Environmental and Planning					
EN1. % household waste recycled	5.29	52.74	2 (Neath PT)	13 (Pembs)	22
EN2. Household refuse collections missed, per 100,000 households	78.86	338.03	2 (Monmouth)	1268 (RCT)	19
EN3. Cost per household of refuse collection	28.17	35.42	14 (Conwy)	53 (Ceredigion)	21
EN4. % householder applications decided within eight weeks	74.59	17.66	30 (Newport)	92 (Bridgend)	22
EN5. The number of appealed planning decisions per 1000 population	0.30	56.66	0.08 (B. Gwent)	0.74 (Monmouth)	22
EN6. % of successful appeals	28.07	51.30	9.30 (Newport)	57.89 (Anglesey)	22
Libraries					
L1. Items issued per head of population	8.23	26.49	7 (Powys)	11 (Monmouth)	22
L2. Visits per head of population	2.55	39.60	1 (Newport)	6 (Carmarthenshire)	21
Finance					
F1. Council tax received as a % of that which should have been received	95.18	2.08	90 (Denbighshire)	98 (Gwynedd)	22
F2. Net cost of tax collection per dwelling	13.59	30.46	2 (Monmouth)	21 (Vale)	21
Protective					
P1. Average number of visits per high and medium risk premises	39.40	52.94	6 (Monmouth)	75 (Denbighshire)	20

C.V. = Coefficient of Variation.

Table 7.3 Descriptive Statistics: Performance Indicators 1997/98

Service	Mean	C.V.	Minimum	Maximum	n
			Education		
E1. % Children under five in LA schools	81.32	14.58	59 (Powys)	100 (Wrexham)	22
E2. % of primary pupils admitted in excess of school capacity	4.65	65.16	1 (Carms)	11 (Caerphilly)	21
E3. % of secondary pupils admitted in excess of school capacity	2.55	191.76	1 (Ceredigion)	22 (Conwy)	15
E4. % of Special Need Statements issued within 18 weeks	59.95	41.40	3 (Carmarthenshire)	87 (Vale)	20
E6. % Mandatory student awards paid on time	58.15	83.80	95 (Cardiff)	99 (Caerphilly)	15
			Social Services		
SS1. Number of physically disabled receiving care, per 1000 adults	4.59	111.11	1.40 (Ceredigion)	19.84 (Carms)	20
SS2. Number of people with learning difficulties receiving care, per 1000 adults	3.11	53.37	0.49 (Gwynedd)	5.72 (Monmouth)	21
SS3. Number of people with mental health problems receiving care, per 1000 adults	3.16	162.66	0.11 (Flintshire)	23.92 (Anglesey)	21
			Housing		
H1. Average number of households in temporary accommodation, per 1000 households	2.76	211.59	0.20 (Merthyr)	14 (Cardiff)	22
H2. Average length of stay in temporary accommodation (weeks)	7.00	57.85	1 (Anglesey)	17 (Monmouth)	22
H3. % of Housing Benefit claims processed in 14 days	78.32	23.49	40 (Denbighshire)	99 (Monmouth)	21

Table 7.3 Continued

Service	Mean	C.V.	Minimum	Maximum	n
H4. Cost of Housing Benefit administration per claimant	52.31	33.07	25 (Wrexham)	74 (Ceredigion)	21
H5. % of tenants owing 13 weeks+ rent	2.54	67.32	1 (Merthyr)	7 (Vale)	22
H6. Average time taken to relet/refurbish dwellings	5.62	40.74	2 (Powys)	12 (B. Gwent)	22
Environmental and Planning					
EN1. % household waste recycled	5.72	58.74	2 (Newport)	13 (Ceredigion)	22
EN2. Household refuse collections missed, per 100,000 households	55.41	234.93	1 (Monmouth)	611 (Bridgend)	19
EN3. Cost per household of refuse collection	33.06	39.29	17 (Monmouth)	69 (Ceredigion)	22
EN4. % householder applications decided within eight weeks	77.68	15.47	36 (Torfaen)	90 (Bridgend)	22
EN5. The number of appealed planning decisions by population	0.30	56.66	0.07 (B. Gwent)	0.71 (Newport)	22
EN6. % of successful appeals	24.86	50.16	6.25 (Pembrokeshire)	46.67 (Ceredigion)	22
Libraries					
L1. Items issued per head of population	7.56	14.02	6 (Flintshire)	10 (Monmouth)	22
L2. Visits per head of population	4.70	22.34	4 (Anglesey)	8 (Denbighshire)	21
Finance					
F1. Council tax received as a % of that which should have been received	95.10	2.24	91 (Cardiff)	98 (Denbighshire)	22
F2. Net cost of collection per dwelling	14.23	28.81	5 (Denbighshire)	20 (Torfaen)	22
Protective					
P1. Average number of visits per high and medium risk premises	37.05	57.94	3 (Merthyr)	85 (Neath PT)	20

C.V. = Coefficient of Variation.

Table 7.4 Descriptive Statistics: Change in Performance 1996/97 – 1997/98

Service	Mean	C.V.	Minimum	Maximum	n
Education					
E1. % Children under five in LA schools	2.20	443.63	−16.30 (Pembs)	24.50 (Conwy)	22
E2. % of primary pupils admitted in excess of school capacity	−0.66	357.57	−6.30 (B. Gwent)	4.40 (Newport)	22
E3. % of secondary pupils admitted in excess of school capacity	0.76	631.57	−3.30 (Newport)	21.80 (Conwy)	22
E4. % of Special Need Statements issued within 18 weeks	16.50	152.61	−16 (Ceredigion)	71 (RCT)	20
E6. % Mandatory student awards paid on time	−34.84	158.40	−2 (Swansea)	68.00 (Conwy)	19
Social Services					
SS1. Number of physically disabled receiving care, per 1000 adults	−0.66	492.00	−15.0 (Ceredigion)	12.08 (Anglesey)	19
SS2. Number of people with learning difficulties receiving care, per 1000 adults	−0.70	407.14	−12.03 (B. Gwent)	2.84 (Powys)	21
SS3. Number of people with mental health problems receiving care, per 1000 adults	−0.84	396.42	−12.91 (Wrexham)	2.80 (Anglesey)	20
Housing					
H1. Average number of households in temporary accommodation, per 1000 households	−0.66	426.34	−17.20 (Flintshire)	8.40 (Newport)	22
H2. Average length of stay in temporary accommodation (weeks)	−0.10	4040.0	−10.00 (Denbighshire)	10.00 (Bridgend)	21
H3. % of Housing Benefit claims processed in 14 days	−3.67	403.81	−41.00 (Flintshire)	27.00 (Newport)	21

Table 7.4 Continued

Service	Mean	C.V.	Minimum	Maximum	n
H4. Cost of Housing Benefit administration per claimant	0.56	3153.5	−34.08(Wrexham)	34.14 (Caerphilly)	20
H5. % of tenants owing 13 weeks+ rent	−0.54	312.46	−10.30 (Denbighshire)	4.10 (Neath)	22
H6. Average time taken to relet/refurbish dwellings	0.58	405.17	−5.00 (Torfaen)	6.00 (Bridgend)	20
Environmental and Planning					
EN1. % household waste recycled	0.44	497.7	−6.30 (Newport)	5.20 (Ceredigion)	22
EN2. Household refuse collections missed, per 100,000 households	23.45	1365.7 18	−28.50 (Merthyr)	603 (Bridgend)	
EN3. Cost per household of refuse collection	4.85	171.13	−5.37 (Bridgend)	24.90 (Carms)	21
EN4. % householder applications decided within eight weeks	3.09	532.00	−35.00 (Torfaen)	57.00 (Newport)	22
EN5. The number of appealed planning decisions by population	0.0	0	−0.25 (Monmouth)	0.20 (Carmarthenshire)	22
EN6. % of successful appeals	−3.20	562.2	−31.11 (Flintshire)	29.05 (Denbighshire)	22
Libraries					
L1. Items issued per head of population	−0.40	91.50	−1.20 (Ceredigion)	0.01 (Swansea)	22
L2. Visits per head of population	2.08	73.56	−1.52 (Conwy)	5.28 (Newport)	20
Finance					
F1. Council tax received as a % of that which should have been received	−0.08	222.92	−3.40 (Carms)	7.90 (Denbighshire)	22
F2. Net cost of collection per dwelling	0.86	538.37	−5.82 (Vale)	16.55 (Monmouth)	21
Protective					
P1. Average number of visits per high and medium risk premises	−3.20	571.87	−0.56 (Denbighshire)	26 (RCT)	19

C.V. = Coefficient of Variation.

Table 7.5 Correlations Between Performance Indicators: Education 1996/97

	E1	E2	E3	E4	E5
E2	0.41	1			
E3	0.13	−0.16	1		
E4	−0.51*	−0.12	0.01	1	
E5	0.21	0.07	−0.02	0.00	1
E6	0.07	0.13	−0.32	−0.46	0.35

Table 7.6 Correlations Between Performance Indicators: Education 1997/98

	E1	E2	E3	E4	E5
E2	0.18	1			
E3	0.04	0.15	1		
E4	−0.19	0.04	−0.06	1	
E5	0.36	0.14	0.31	−0.23	1
E6	0.04	0.10	0.19	−0.11	−0.37

Table 7.7 Correlations Between Performance Indicators: Education Change 1996/97–1997/98

	E1	E2	E3	E4	E5
E2	0.34	1			
E3	0.17	−0.23	1		
E4	−0.44	−0.18	−0.32	1	
E5	0.44	0.04	0.34	−0.21	1
E6	0.30	0.13	0.13	−0.32	0.03

Table 7.8 Correlations Between Performance Indicators: Social Services 1996/97

	SS1	SS2
SS2	0.15	1
SS3	0.18	0.23

Table 7.9 Correlations Between Performance Indicators: Social Services 1997/98

	SS1	SS2
SS2	0.18	1
SS3	0.58*	0.19

Key to significance levels in Tables 7.5–7.18: + ≤ 0.10   * ≤ 0.05   ** ≤ 0.01.

Table 7.10 Correlations Between Performance Indicators: Social Services Change 1996/97–1997/98

	SS1	SS2
SS2	0.13	1
SS3	0.28	0.16

Table 7.11 Correlations Between Performance Indicators: Housing 1996/97

	H1	H2	H3	H4	H5
H2	0.00	1			
H3	0.11	–0.08	1		
H4	–0.04	–0.03	0.27	1	
H5	0.16	–0.34	–0.03	0.15	1
H6	–0.08	–0.00	–0.57*	–0.43+	0.13

Table 7.12 Correlations Between Performance Indicators: Housing 1997/98

	H1	H2	H3	H4	H5
H2	–0.10	1			
H3	–0.06	–0.01	1		
H4	–0.16	0.23	–0.33	1	
H5	0.14	–0.14	0.08	0.04	1
H6	0.06	–0.11	–0.06	0.26	0.26

Table 7.13 Correlations Between Performance Indicators: Housing Change 1996/97– 1997/98

	H1	H2	H3	H4	H5
H2	0.08	1			
H3	0.64	–0.22	1		
H4	–0.22	0.24	–0.13	1	
H5	–0.09	–0.11	0.01	–0.19	1
H6	0.28	0.14	0.11	–0.04	–0.07

Key to significance levels in Tables 7.5–7.18: + ≤ 0.10   * ≤ 0.05   ** ≤ 0.01.

Table 7.14 Correlations Between Performance Indicators: Environmental 1996/97

	EN1	EN2	EN3	EN4	EN5
HN2	−0.30	1			
EN3	−0.10	−0.03	1		
EN4	0.08	0.16	−0.22	1	
EN5	0.03	0.15	0.12	0.63*	1
EN6	0.01	0.10	0.05	−0.15	−0.08

Table 7.15 Correlations Between Performance Indicators: Environmental 1997/98

	EN1	EN2	EN3	EN4	EN5
EN2	−0.02	1			
EN3	−0.41	0.25	1		
EN4	−0.16	−0.12	0.17	1	
EN5	−0.05	−0.01	0.06	0.67**	1
EN6	0.06	−0.09	0.03	−0.15	−0.27

Table 7.16 Correlations Between Performance Indicators: Environmental Change 1996/97–1997/98

	EN1	EN2	EN3	EN4	EN5
EN2	−0.01	1			
EN3	0.03	0.09	1		
EN4	−0.49+	−0.09	0.19	1	
EN5	−0.31	−0.01	0.25	0.22	1
EN6	0.12	−0.01	−0.38	−0.13	−0.09

Table 7.17 Correlations Between Performance Indicators: Leisure Services 1996/97, 1997/98 and Change 1996/97 – 1997/98

	Coefficients
L1 WITH L2, 1996/7	0.54*
L1 WITH L2, 1997/98	0.25
L1 WITH L2, 1996/7–1997/8	−0.03

Key to significance levels in Tables 7.5–7.18: + ≤ 0.10   * ≤ 0.05   ** ≤ 0.01.

Table 7.18 Correlations Between Performance Indicators: Financial Services 1996/97, 1997/98 and Change 1996/97 – 1997/98

	Coefficients
F1 WITH F2, 1996/7	0.02
F1 WITH F2, 1997/98	−0.26
F1 WITH F2, 1996/7–1997/8	0.11

(a) *Financial Resources* Since the 1970s there has been a widespread concern in western nations to limit levels of taxation and public expenditure. A popular catchphrase of anti-statists has been that 'problems are not solved by throwing money at them'. Yet, if money is not a sufficient condition of good public services, it may be still a necessary condition. Organizations that are wealthy are likely to find it easier to produce a greater quantity of services, to achieve higher quality, and to deliver services speedily. Moreover, some political parties may make extravagant promises in the knowledge that money is available to keep them, whereas others are more circumspect in their commitments because they know that funds are scarce. Thus it is necessary to take account of variations in financial prosperity when assessing the relationship between manifesto pledges and service performance.

The level of financial resources in a local authority can be operationalized through the total net expenditure on services. We measure this variable as the *estimated* level of spending prior to the start of the financial year in which the performance variables are measured. This lag structure should remove any potential simultaneity bias in the estimated relationship between resources and service performance. For example, decisions to provide wide service coverage or high levels of administrative effectiveness are likely to entail more expenditure. Any reciprocal effects that remain will in any case be absent from the analysis of the impact of financial resources on performance *change* in later years. On the basis of the arguments above, and prior empirical evidence (Dean & Peroff, 1977), we hypothesize a positive relationship between expenditure and service performance.

(b) *Party Control.* The measurement of this variable is based upon whether the authority was Labour-controlled from the 1995 elections. Empirical tests of the ideological model of party behaviour have focused on general policy priorities, not specific election pledges (Dye, 1966; Castles, 1982; Sharpe & Newton, 1984). Typically, the procedure has been to assess the link between expenditure, in a variety of policy areas (particularly welfare, education and health) and party political control. It is normally anticipated that left-wing control will result in an increase in expenditures on redistributive policies, whilst decreases in spending on welfare are expected from right-wing parties in government (Hofferbert & Budge, 1992).

At a national level, there is some evidence to support the relationship between ideology and expenditure, with socialist governments linked to welfare spending and Conservative governments prioritising defence spending (Budge & Hofferbert, 1990). Similarly, at a local level in the U.K., Labour Party control is generally associated with higher spending, especially on services such as education, housing and social services (Boyne, 1996). Although there is no prior empirical evidence in this issue, it seems plausible that Labour councils do not see extra spending as an end in itself. Rather, the intention is to promote better public services. We therefore hypothesize that Labour control is positively related to service performance.

Summary of the Model of Performance Variations

Our multivariate model of performance levels in 1996/97 and 1997/98 contains three explanatory variables: the proportion of manifesto pledges on the service in question, the level of financial resources available for all services provided by a local authority, and the political control of the council. The relationship between these three variables is shown in Table 7.19. The links between them are generally weak and insignificant, so the statistical estimates should not be clouded by multicollinearity. The model for performance changes over time contains an extra 'explanatory' variable: the level of service performance in the 'base year' 1996/97. This is not an additional substantive explanatory variable as such. Rather, its inclusion in the model simply accommodates a statistical artefact: the scope for improvement is greatest where performance is already

Table 7.19 Relationships between Explanatory Variables

Election Pledges	Expenditure	Labour Party Control
Education Election Pledges	–0.24	0.20
Social Services Election Pledges	–0.16	0.28
Housing Election Pledges	0.01	0.20
Highways Election Pledges	–0.11	0.33
Environmental Election Pledges	0.10	0.28
Libraries Election Pledges	–0.05	0.30
Finance Election Pledges	–0.19	0.19
Protective Election Pledges	–0.17	0.40
Labour Party Control	–0.43*	–

weak, and smallest where it is already strong. The latter phenomenon is especially pronounced for variables that have an absolute ceiling (e.g. indicators E4, E6, H3, H5, EN4 and F1 in Table 7.2).

STATISTICAL RESULTS

The variables were entered in an OLS multiple regression equation, the results of which are shown in Table 7.20 and Table 7.21. These tables show the level of statistical explanation produced by the model (R^2 and Adjusted R^2) and the coefficients for explanatory variables that are significant at the 0.10 level or better.

Performance Levels

Taken together, the three explanatory variables generally provide a weak and insignificant explanation of variations in service performance. The average R^2 in 1996/97 is 0.16, and only five of the figures are statistically significant (for dependent variables E3, E6, H4, EN3, and EN5 – Table 7.20). The performance of the model is slightly better in 1997/98: the average R^2 rises to 20%, but the level of statistical explanation is again significant in only five cases (variables H6, EN3, EN5, EN6, and F1). Thus, the model 'works' in both years for only two measures of performance (EN3 – the cost per household of refuse collection, and EN5 – the percentage of planning decisions that are appealed against).

The weakness and instability of the results is also reflected in the coefficients for the individual explanatory variables. The main variable of interest here, manifesto pledges, is significant only twice in the 1996/97 equations. Furthermore, it has the 'wrong' sign: a higher proportion of pledges on environmental services is associated with worse performance (more planning appeals and more successful appeals). Although the pledges variable is significant in four cases in 1997/98, the implications for mandate theory are mixed. Manifesto pledges on education are associated with better performance on indicator E2 (fewer primary pupils are taught in overcrowded schools), and pledges on social services are positively related to indicator SS1 (more physically disabled people receive council care). By contrast, pledges on housing are linked with poorer performance: slower processing of claims for housing benefit (indicator H3) and more tenants in serious rent arrears (indicator H5). Moreover, as in 1996/97, pledges on environmental services are associated with a higher rate of planning appeals. Thus the coefficient for manifesto pledges is significant in only seven of fifty tests, and five of these coefficients flatly contradict mandate theory.

Table 7.20 Multivariate Regression Results for Performance Levels in 1996/97 and 1997/98

Variable	Coefficients 1996/97		Significant Variables	Coefficients 1997/98		Significant Variables
	R^2	Adj R^2		R^2	Adj R^2	
Education						
E1. % Children under five in LA schools	0.23	0.10		-0.18	0.04	
E2. % of primary pupils admitted in excess of school capacity	0.09	-0.07		0.27	0.15	PL -0.45*
E3. % of secondary pupils admitted in excess of school capacity	0.53**	0.43	Exp-0.39* Lbc 0.50*	0.22	0.09	Exp -0.40+ Lbc 0.42*
E4. Speed of issue of special needs statements	0.07	-0.09		0.09	-0.07	
E6. Mandatory student awards paid on time	0.33+	0.21	Exp 0.62* Lbc 0.47*	0.14	-0.02	
Social Services						
SS1. Number of physically disabled receiving care, per 1000 adults	0.11	-0.04		0.18	0.05	PL 0.43+
SS2. Number of people with learning difficulties receiving care, per 1000 adults	0.22	0.09	Lbc 0.42+	0.13	-0.03	
SS3. Number of people with mental health problems receiving care, per 1000 adults	0.03	-0.12		0.10	-0.05	
Housing						
H1. Average number of households in temporary accommodation, per 1000 households	0.03	-0.13		0.10	-0.04	
H2. Average length of stay in temporary accommodation	0.07	-0.09		0.04	-0.12	
H3. % of Housing Benefit claims processed in 14 days	0.09	-0.06		0.14	-0.00	PL -0.38+

Pledges and Performance 195

Table 7.20 Continued

Variable	Coefficients 1996/97 R²	Adj R²	Significant Variables	Coefficients 1997/98 R²	Adj R²	Significant Variables
H4. Cost of Housing Benefit administration per claimant	0.41*	0.31	Exp 0.57*	0.20	0.06	
H5. % of tenants owing 13 weeks+ rent	0.09	−0.06		0.27	0.15	PL 0.40+
H6. Average time taken to relet/refurbish dwellings	0.11	−0.03		0.31+	0.19	Lbc 0.50
Environmental and Planning						
EN1. % household waste recycled	0.19	0.06		0.18	0.04	Lbc −0.45+
EN2. Household refuse collections missed, per 100,000 households	0.03	−0.17		0.17	0.03	
EN3. Cost per household of refuse collection	0.39*	0.29	Exp 0.62*	0.47**	0.38	Exp 0.56*
EN4. % householder applications decided within eight weeks	0.00	−0.16		0.08	−0.07	
EN5. The number of appealed planning decisions by population	0.38+	0.23	PL 0.49*	0.44*	0.34	PL 0.51*
EN6. % of successful appeals	0.26	−0.14	PL 0.51*	0.38*	0.27	
Libraries						
L1. Items issued per head of population	0.01	−0.15		0.08	−0.07	
L2. Visits per head of population	0.04	−0.11		0.02	−0.13	
Finance						
F1. Council tax received as a % of that which should have been received	0.06	−0.10		0.41*	0.32	Exp-0.51* Lbc-0.68**
F2. Net cost of collection per dwelling	0.10	−0.06		0.27	0.15	Lbc 0.54*
Protective						
P1. Average number of visits per high and medium risk premises	0.15	−0.01	0.14	−0.02		

Key for Tables 7.20–7.21:   Exp = Total Expenditure per Capita 1996/97   Lbc = Labour Party Control   PL = Election Pledge   Base Pl = Pl at 1996/97 level
Significance levels: + ≤ 0.10   * ≤ 0.05   ** ≤ 0.01   *** ≤ 0.001

Table 7.21 Multivariate Regression Results for Performance Change Between 1996/97 and 1997/98

Variable	R^2	Coefficients Change 1996–97 including PI Base variable Adj R^2	Significant Variables
Education			
E1. % Children under five in LA schools	0.28	0.12	Base PI −0.58*
E2. % of primary pupils admitted in excess of school capacity	0.86**	0.83	Base PI −0.93*** PL −0.20*
E3. % of secondary pupils admitted in excess of school capacity	0.88***	0.85	Base PI −0.88***
E4. Speed of issue of special needs statements	0.44**	0.29	Base PI −0.47*
E6. Mandatory student awards paid on time	0.99***	0.99	Base PI −1.0***
Social Services			
SS1. Number of physically disabled receiving care, per 1000 adults	0.77***	0.71	Base PI −0.82*** PL 0.23+
SS2. Number of people with learning difficulties receiving care, per 1000 adults	0.74***	0.67	Base PI −0.88***
SS3. Number of people with mental health problems receiving care, per 1000 adults	0.70***	0.62	Base PI −0.79*** Exp 0.26
Housing			
H1. Average number of households in temporary accommodation, per 1000 households	0.70***	0.62	Exp 0.26+ Base PI −0.79***
H2. Average length of stay in temporary accommodation	0.37+	0.22	Base PI −0.57+
H3. % of Housing Benefit claims processed in 14 days	0.21	0.02	Base PI −0.39+
H4. Cost of Housing Benefit administration per claimant	0.40*	0.26	Base PI −0.64*

Table 7.21 Continued

Variable	Coefficients Change 1996–97 including PI Base variable		Significant Variables
	R^2	Adj R^2	
H5. % of tenants owing 13 weeks+ rent	0.79***	0.74	PL 0.22+ Base PI –0.80***
H6. Average time taken to relet/refurbish dwellings	0.45*	0.32	Base PI –0.58* Lbc 0.42+
Environmental and Planning			
EN1. % household waste recycled	0.11	–0.10	
EN2. Household refuse collections missed, per 100,000 households	0.85***	0.82	Base PI –0.92***
EN3. Cost per household of refuse collection	0.28	0.10	
EN4. % householder applications decided within eight weeks	0.48*	0.36	Base PI –0.67**
EN5. Number of appealed planning decisions by population	0.38+	0.23	Base PI –0.44*
EN6. % of successful appeals	0.54**	0.43	Base PI –0.70**
Libraries			
L1. Items issued per head of population	0.44*	0.30	Base PI –0.66*
L2. Visits per head of population	0.52*	0.39	Base PI –0.71**
Finance			
F1. Council tax received as a % of that which should have been received	0.54**	0.44	Exp –0.41* Lbc –0.60** Base PI –0.59**
F2. Net cost of collection per dwelling	0.52*	0.40	Base PI –0.66** Lbc 0.46*
Protective			
P1. Average number of visits per high and medium risk premises	0.24	0.04	

The results for financial resources and party control are a little better. The prior level of expenditure is associated with higher service performance in three cases (indicators E3 and E6 in 1996/97, and E3 in 1997/98), and lower performance in four cases (indicators H4 and EN3 in 1996/97, and EN3 and F1 in 1997/98). The latter results are consistent with the view that higher spending councils are less efficient. The party politics variable is significant in eight of the fifty or so tests, but seldom in the hypothesized direction. Labour control is associated with higher performance on indicators E6 and SS2 in 1996/97, which supports the view that the party is especially committed to the provision of education and social services. However, Labour control is associated with poorer performance on indicator E3 (in both years), and indicators H6, EN1 and F1. Thus, the link established in previous research between Labour dominance and spending does not appear to be translated into a positive effect on service standards.

Performance Change

The level of statistical explanation produced by the performance change model is substantially higher than that for performance levels. The average R^2 is as high as 0.54, and is significant in all but five cases (indicators E1, H3, EN1, EN3 and P1). This statistical success is largely attributable to the strong negative relationship between the base measure of performance (as reflected, for example, in the results for indicator E6 in Table 7.21). Otherwise, the results are much as before. The manifesto pledges variable is significantly associated with an improvement in performance in two cases (indicators E2 and SS1) but with a deteriorated performance in one case (indicator H5). Financial resources are linked with better performance in one test (indicator SS3) but with a worse performance in two tests (indicators H1 and F1). All three significant coefficients for Labour control imply a decline in performance on indicators of housing relet times (H6) and council tax collection (F1 and F2).

CONCLUSION

The impact of manifesto pledges on public spending and policies has been examined widely at the level of national governments, but has been virtually ignored at the local level. Furthermore, there has been no analysis, either nationally or locally, of the relationship between manifesto commitments and service performance in the public sector. In this chapter we have provided the first empirical analysis of the issue, by examining the relationship between pledges and performance in Welsh local government.

Mandate theory suggests that democratic governments are responsive to public preferences because they deliver on their promises. In other words, the standard of public services should be at least partly predictable on the basis of the election pledges of the ruling party. Our statistical results provide very little support for this view of the political process. Neither levels of performance nor changes in performance over time appear to be influenced much by the priority attached to different services in party manifestos. A pledge by a ruling party to pay special attention to a service area is as likely to be followed by deterioration as improvement in standards.

At this point it is important to emphasize the preliminary and exploratory nature of our research. We have analyzed only one group of local authorities over a short time-period. Indeed, it may be that data over the whole four-year electoral cycle (rather than simply the first two years) will eventually show different results. Also, service performance is clearly a complex phenomenon that requires a sophisticated explanatory model. It may be that if measures of manifesto pledges are included in such a model then their significance may emerge. For the present, however, we must conclude that in these authorities, in this time-period, and on the basis of the variable in our model, the evidence contradicts the hypothesized positive relationship between manifesto pledges and service performance.

REFERENCES

Ashworth, R. E. (2000a). *Promises and Performance: An Empirical Analysis of the Relationship Between Election Pledges and Local Government Policy-Making.* Unpublished Ph.D. Thesis, Cardiff University.

Ashworth, R. E. (2000b). Party Manifestos and Local Accountability: A Content Analysis of Local Election Pledges in Wales. *Local Government Studies* forthcoming.

Berelson, B. (1952). *Content Analysis in Communications Research.* New York: Free Press.

Boyne, G. A. (1996). Assessing Party Effects on Local Policies: a Quarter Century of Progress or Eternal Recurrence. *Political Studies 44,* 232–252.

Boyne, G. A. (1997). Comparing the Performance of Local Authorities: An Evaluation of Audit Commission Indicators. *Local Government Studies* 23(4), 17–43.

Budge, I., & Hofferbert, R. I. (1990). Mandates and Policy Outputs: US Party Platforms and Federal Expenditures. *American Political Science Review 84,* 113–131.

Budge, I., Robertson, D., & Hearl, D. (Eds.). (1987). *Ideology, Strategy and Party Change,* Cambridge: Cambridge University Press.

Cameron, K. (1986). Effectiveness as Paradox: Consensus and Conflict in Perceptions of Organizational Effectiveness. *Management Science 32,* 539–553.

Carter, N., Day, P., & Klein, R. (1992). *How Organizations Measure Success.* London: Routledge.

Castles, F. G. (Ed.). (1982). *The Impact of Parties: Politics and Policies in Democratic Capitalist States*. Beverly Hills: California: Sage.

Connolly, T., Conlon, E., & Deutsch, S. (1980). Organizational Effectiveness: A Multiple Constituency Approach. *Academy of Management Review 5*, 211–217.

Cutler, T., & Waine, C. (1997). *Managing the Welfare State*. Oxford, Berg.

Dean, G., & Peroff, K. (1977). The Spending-Service Cliché: An Empirical Re-Evaluation. *American Politics Quarterly 5*, 501–516.

Durant, R. (1999). The Political Economy of Results-Oriented Management in the Neo-Administrative State. *American Review of Public Administration 29*, 307–331.

Dye, T. (1966). *Politics, Economics and the Public*. Chicago: Rand-McNally.

Elling, R. C. (1979). State Party Performance and State Legislative Performance: A Comparative Analysis. *American Journal of Political Science 23*, 383–405.

Finer, S. E. (1975). Manifesto Moonshine. *New Society* 13 November: 379–381.

Fudge, C. (1981). Winning Elections and Taking Control: The Formulation and Implementation of a Local Manifesto. In: S. Barrett, & C. Fudge (Eds.), *Policy and Action: Essays on the Implementation of Public Policy*, (123–142). London: Methuen.

Game, C., & Skelcher, C. (1983). Manifestos and Other Manifestations of Local Party Politics – The Spread of Party Politics Since Reorganization. *Local Government Studies (Annual Review)* July/August, 29–33.

Ginsberg, B. (1976). Elections and Public Policy. *American Political Science Review 70*, 7–27.

Gyford, J. Leach, S., & Game, C. (1989). *The Changing Politics of Local Government*. London: Unwin Hyman.

Hofferbert, R .I., & Klingemann, H. (1990). The Policy Impact of Party Programmes and Government Declarations in the Federal Republic of Germany. *European Journal of Political Research 18*, 277–304.

Hofferbert, R. I., & Budge, I. (1992). The Party Mandate and the Westminster Model: Election Programmes and Government Spending in Britain, 1945–1985. *British Journal of Political Science 22*, 151–182.

Hogwood, B. W., & Gunn, L. A. (1984). *Policy Analysis for the Real World*. Oxford: Oxford University Press.

Hood, C., Scott, C., James, O., & Travers, T. (1998). *Regulation Inside Government*. Oxford: Oxford University Press.

Kalogeropoulou, E. (1989). Election Promises and Government Performances in Greece: PASOKs Fulfilment of its 1981 Election Pledges. *European Journal of Political Research 17*, 289–311.

Kimberley, J., Norling, F., & Weiss, J. (1983). Pondering the Performance Puzzle. In: R. Hall, & R. Quinn (Eds.), *Organizational Theory and Public Policy*, (249–264). London, Sage.

Krippendorff, K. (1980). *Content Analysis: An Introduction to its Methodology*. London: Sage.

Ott, J., & Shafritz, J. (1994). Toward a Definition of Organizational Incompetence: A Neglected Variable in Organizational Theory. *Public Administration Review 54*, 370–377.

Petry, F. (1988). The Policy Impact of Canadian Party Programs: Public Expenditure Growth and Contagion from the Left. *Canadian Public Policy 14*, 376–389.

Pomper, G., & Lederman, S. S. (1980). *Elections in America*. New York: Longman.

Quinn, R., & Rohrbaugh, J. (1981). A Competing Values Approach to Organizational Effectiveness. *Public Productivity Review 5*, 122–140.

Quinn, R., & Rohrbaugh, J. (1983). A Spatial Model of Effectiveness Criteria: Towards A Competing Values Approach to Organizational Analysis. *Management Science 29*, 363–377.

Rallings, C. (1987). The Influence of Election Programmes: Britain and Canada 1945–1979. In: I. Budge, D. Robertson, & D. Hearl (Eds.), *Ideology, Strategy and Party Change*, (1–14). Cambridge: Cambridge University Press.

Rallings, C., & Thrasher, M. (1997). *Local Elections in Britain*. London: Routledge.

Robertson, D. (1976). *A Theory of Party Competition*. London: Macmillan.

Rohrbaugh, J. (1983). The Competing Values Approach. In: R. Hall & R. Quinn (Eds.), *Organizational Theory and Public Policy*, (265–280). London: Sage.

Rose, R. (1980). *Do Parties Make a Difference?*. London: Macmillan.

Royed, T. (1996). Testing the Mandate Model in Britain and the United States: Evidence from the Reagan and Thatcher Eras. *British Journal of Political Science 26*, 45–80.

Sharpe, L. J., & Newton, K. (1984). *Does Politics Matter?*. Oxford: Clarendon.

Steers, R. (1975). Problems in The Measurement of Organizational Effectiveness. *Administrative Science Quarterly 20*, 546–558.

Stewart, J., & Walsh, K. (1994). Performance Measurement When Performance Can Never Be Finally Defined. *Public Money and Management, 14*, 45–49.

Stoker, G. (1991). *The Politics of Local Government*. London: Macmillan.

Strom, K., & Leipart, J. Y. (1989). Ideology, Strategy and Party Competition in Post-War Norway. *European Journal of Political Research 17*, 289–311.

Topf, R. (1994). Party Manifestos. In: A. Heath, R. Jowell, & J. Curtice (Eds.), *Labours Last Chance*, (149–172). Aldershot: Dartmouth.

Weber, R. P. (1990). *Basic Content Analysis*. London: Sage.

Wilson, D., & Game, C. (1998). *Local Government in the United Kingdom*. London: Macmillan.

8. URBAN POLICY MAKING IN GERMANY: THE IMPACT OF PARTIES ON MUNICIPAL BUDGETS AND EMPLOYMENT

Volker Kunz

Since the middle of the eighteenth-century, institutions of local self-administration in Germany have served important public functions. Today, municipalities have the responsibility to regulate all affairs in the local community within an established legal framework. Hence, contrary to the constitutional tradition prevailing in the Weimar Republic, local government is not institutionally separated from the state apparatus. According to Article 28 of the German Basic Law, local government is an integral part of the democratic political system. Allied to this formal responsibility, local political decisions have considerable impact on the spatial distribution and possible uses of economic, social and cultural public infrastructure. Municipalities are responsible for about 60% of all public investment. They implement about 80% of federal and state laws, so they are the public institutions with the most intensive contact with the citizenry. To accomplish their various tasks, municipalities invest large sums and employ numerous workers, so their personnel expenditure is high. Local government also plays an important role in promoting economic growth, as well as in providing collective goods.

One key focus of research on local government is the impact of local institutions and the process of authoritative decision-making on policy. For theories of democracy and human welfare this issue is of great significance. If control of political institutions through political parties, the intensity of local political competition or other characteristics of the local political system, have no influence on local policy, then the concept of a competitive democracy is questioned (Dahl, 1971; Lijphart, 1984; Schmidt, 1996). On the other hand, local democracy seems to face manifold problems as a result of a spiral of rising expectations and a deep fiscal crisis. This was seen in the 1980s and 1990s, when efforts to consolidate Germany's local government finances, following the U.S. and U.K. experiences with urban fiscal crisis (e.g. Alcaly & Mermelstein, 1977; Pammer, 1990), met with an increasing employment crisis and a regional differentiation in social problems. These factors led to a 'polarization of German cities' (Haeussermann & Siebel, 1986; Haeussermann et al., 1995). Set against pressure for fiscal restraint, local budgeting strategies, especially investment projects and the personnel policies of municipalities and local public enterprises, provide an important opportunity to support the local labor market and reduce the welfare needs of the local population. Alongside this, political actors feel pressured to cope with social and fiscal problems in a way that corresponds with citizen expectations (Holtmann, 1992, p. 19). The assumption underlying this chapter is that this leads to vehement debate at the national and local levels between two main positions: a demand-oriented call for compensatory increases in expenditure and staffing, which should be financed by additional taxes, charges or credits; and, a supply-oriented call for cuts in expenditure and staffing levels, which should lower taxes. The anticipation is that this political debate will manifest itself in dissimilar local policy decisions.

This article analyzes the impact of party dominance on local budget and employment decisions. Five policy indicators are examined for all 87 county-free cities in West Germany during the 1980s. With their special position in the system of public administration, these 'city counties' combine administrative functions that are normally divided between counties and municipalities (for details see Gunlicks, 1986). The selection of these cities and the limitation of the time-period have their roots in the availability of reliable data.

Assuming that political parties try to realize their political programs at a local level, the success of their efforts should be revealed in local financial decisions. The hypothesis that political parties promote different financial decisions is based on the presupposition that local policy decisions in Germany are largely shaped by political parties. It also presupposes a competitive style of local politics that goes beyond Downs's (1957) purely office-seeking behavior. The responsiveness of local policy-makers to the perceived demands of the majority of the electorate is assumed to be filtered through ideological dispositions. Party affiliation and associated ideological dispositions amongst local policy-makers should be important determinants of policy preferences. The correlation between policy-makers' preferences and the perceived demands of the majority of voters might be rather weak. These assumptions will be discussed first in the next section.

THE IMPACT OF PARTIES: MICRO-LEVEL EVIDENCE

The dominance of a legal-institutional approach in German research on local government and politics, and a prevalent rejection of quantitative methods (e.g. Puettner, 1982; Heinelt, 1991, p. 266), means there has been little empirical research on how political variables affect local policy-making in Germany (among the rare exceptions are Fried, 1976; Bennett, 1985; Rickards, 1985; Gruener et al., 1988; Bothe, 1989; Kunz & Zapf-Schramm, 1989; Gabriel et al., 1990). As far as empirical work on the causes and consequences of local policy decisions are concerned, descriptive accounts and qualitative case-studies of decision processes, power structures and policy impacts predominate (e.g. Winkler-Haupt, 1988; Gau, 1990; Jaedicke et al., 1990; McGovern, 1997). Hence, irrespective of a strong tradition of local policy output research in the international science community (e.g. Sharpe & Newton, 1984; Hoggart, 1989; Boyne, 1992), recent broadly based analyses of the 'impact of parties' do not exist for German cities (the only exception is Kunz, 2000). Those studies that do offer some insight on this issue are outdated, rest on a small sample of local units or come from a regional science stable with a strong bias toward the issue of fiscal equalization (e.g. Mielke, 1985; Kunz & Gabriel, 1992; Miera, 1994; Pohlan, 1997). As a consequence there is a one-sided emphasis on socio-economic and fiscal determinants of local policy. This corresponds to the traditional view of German local government as nonpolitical, conflict-free and an ideologically neutral regulator of administrative tasks (see Lehmbruch, 1979; Gabriel, 1984, 2000).

Yet the local political system in Germany is shaped by political parties, with competition mainly structured by left-right differences. Especially in county-free cities, politics is dominated by the CDU/CSU and the SPD. The average share of posts for these parties on local councils in the 1980s and 1990s was about 90%. This left-right divide will form the core of the analysis undertaken here on municipal budget and employment policies.[1]

The starting point for this analysis is to indicate the invalidity of the standard assumption of the economic theory of democracy; namely, that "... parties formulate policies in order to win elections, rather than win elections in order to formulate policies" (Downs, 1957, p. 28). A pure strategy of maximizing votes implies that parties hold contradictory positions simultaneously in order to meet as many interests as possible. It additionally suggests that parties orient their positions toward the majority of the electorate for strategic reasons, so their platforms vary marginally from each other (Wright, 1971; Klingemann et al., 1994; Klingemann & Volkens, 1997). Individual level data on councilors' ideological dispositions, party affiliation, selected issue preferences and perceived preferences of the electorate, contradict this position. These data are available from local surveys in six county-free cities in 1991 and 1993.[2] As Table 8.1 shows, for councilors, ideological orientation[3] and party-membership correlate highly (r = 0.75), with left orientations and SPD membership being allied to a preference for increased total, personnel and investment expenditure. The tendency to favor increased personnel expenditure is especially marked for council members of the SPD or those who place themselves on the left of the left-right continuum (r = −0.50 and −5). This orientation corresponds with programs and platforms offered to the electorate by the SPD. Positive expenditure preferences for social services and social infrastructure are also recorded with this political orientation (see Table 8.1). In sum, contrary to the economic theory of democracy, there are systematic, consistent party differences in socio-cultural values (e.g. Frank, 1990, p. 249). The CDU/CSU is supportive of a limited role for local government in social life. Their electoral platforms reveal a disposition toward the market economy and the principle of subsidiarity in providing public goods. In their view, the main purpose of government is to improve conditions for private economic activities. It follows that the local traffic system, garbage disposal, especially for industrial activities, and investment activities, belong to fields where rightist parties in Germany are most likely to want increased spending (see Reuter, 1976, p. 23; Knemeyer & Jahndel, 1991, p. 31).

At least partially, this assumption corresponds with councilor attitudes. Those on the political left do have a tendency to favor higher investment expenditure than rightist councilors, but the relationship is a weak one, especially compared to the stronger preference for social services spending amongst those on the left (Table 8.1). Moreover, with respect to investment in roads and parking lot construction, policy-makers who place themselves on the political right are most favorably disposed toward greater spending (r = 0.49 for parties and r = 0.62 for councilors' self-assessed left-right orientations). These policies can be regarded as integral parts of a conservative economic doctrine that is primarily oriented towards private economic interests. This assessment corresponds with rightist councilors' stronger preference for spending on local economic development. Alongside economic issues, the rightist policy profile also favors spending on public security. Accordingly, higher expenditure on public order is most strongly supported by councilors adhering to rightist values (Table 8.1).[4]

Table 8.1 Links of Left-Right Orientation and Party Affiliation with Spending and Revenue Preferences and Perceptions of Social Problems

Intercorrelations with	Left-right orientation (1 = left, 11 = right)		Party affiliation (1 = SPD, 2 = CDU/CSU)	
	Correlation	N	Correlation	N
Ideology (1 = left, 11 = right)	1.00	200	0.75	141
General spending preferences *				
Total expenditure	−0.39	187	−0.35	144
Personnel expenditure	−0.50	188	−0.55	143
Investment expenditure	−0.18	187	−0.25	142
Expenditure in specific policies *				
Social Services	−0.55	189	−0.60	143
General and youth welfare	−0.55	187	−0.55	143
Public health	−0.22	188	−0.22	143
Schools	−0.13	186	−0.05	141
Sports and leisure facilities	−0.05	188	−0.16	132
Parks and recreation areas	−0.15	189	−0.19	143
Culture	−0.20	187	−0.19	141
Urban development/ renewal	−0.08	189	−0.11	144
Sewage and garbage disposal	−0.14	188	−0.18	144
Street and parking lot construction	0.62	181	0.49	139
Local economic development	0.24	189	0.20	144
Public order and security	0.46	188	0.37	144
Revenue preferences **				
(0 = not mentioned, 1 = mentioned)				
Reduce tax rate to the business tax	0.24	190	0.25	146
Reduce tax rate to the property tax	0.12	190	0.11	146
Reduce charges	0.00	190	0.01	146
Reduce debt	−0.04	190	−0.05	146
Problem perception***				
(1 = least important, 5 = most important)				
Unemployment	−0.25	131	−0.36	107
Rising service demands from citizens	0.23	131	0.13	105

Descriptive values: N-SPD = 82, N-CDU/CSU = 64; mean left-right-orientation = 5.22, standard deviation = 2.71. Correlations equal or higher |0.20| are in bold. QUESTION WORDING: * 'Please think about the following policy fields. What is your opinion on the level of local public expenditure in your city'? The response alternatives were: large cuts (1), some cuts (2), keep the same level of spending (3), some increase (4), large increase (5). ** 'If your government were given an increase in general revenue sharing equal to 20% of total local expenditure, how would you like to see these funds used'? *** 'In the last three years, how important have each of the following problems been for your city's finances'?

Equivalent indicators of policy-makers' preferences in revenue policies are not available (the only revenue question asked focused on the special case of higher tier grant assistance). Yet the often-assumed effects of left-right orientation and party affiliation on revenue preferences are partially supported by the data. Although the relationship is not strong, those on the political right are more in favor of reducing tax rates (*Hebesatz*), especially business taxes, when grants by the state are raised, whereas those on the left are not (r values for business tax rates are 0.24 and 0.25; for property tax rates 0.12 and 0.11). For service charges and debt-policies no left-right distinction was found.

Overall, an ideologically neutral, non-partisan style of local politicis is not observed in the six cities covered. Not only is the national political system shaped by political parties and left-right conflict, but so are local political systems. Moreover, left-right orientations and political party affiliations are important determinants of councilor preferences for most of the issues under scrutiny. The often-stated assertion that local politics in Germany is not strongly influenced by ideological factors is not empirically validated. This applies even to councilor perceptions of societal problems and their influence on municipal finances. Thus, policy-makers on the right perceive a greater influence of rising citizen expectations on the financial situation of their city than those on the left (r = 0.23 and 0.13), who see city finances coming under pressure more on account of unemployment (r = –0.25 and –0.36).[5] This pattern corresponds to the neo-conservative critique of the welfare state.

This relationship between councilor issue preferences and perceived demands from the electorate also contradicts ideas about parties adopting 'pure' vote-maximizing strategies. In a more detailed survey of the four cities of Bamberg, Bonn, Ludwighafen and Wiesbaden, councilors were asked to assess the majority of voters' spending preferences in 21 policy areas (see the items in Table 8.2).[6] On the basis of the mean scores for these items and councilors spending attitudes, respondent preferences and their perceptions of public demands were turned into a rank order. A comparison between the ranks of the two lists of items answers the question of whether representatives behave as vote-maximizers. According to the median voter model, the (perceived) demands of the majority of the electorate should be the main factor influencing the spending preferences of local councilors (Bergstrom & Goodman, 1973). However, the picture given by the rank order lists does not confirm this assertion (Table 8.2). The rank-correlation between the (average) spending preferences of local policy-makers on the one hand and perceived preferences of the electorate on the other is far from strong ($r_s = 0.27$).

Councilor preferences and their perceptions of public demands were also determined separately for CDU/CSU and SPD representatives. As Table 8.3

Table 8.2 Spending Preferences of Politicians and Politicians' Assessments of Electoral Preferences

	Spending preferences of politicians*			Perceived preferences of the elctorate**		
Item	mean	std. dev.	rank	mean	std. dev.	rank
Public transportation	4.38	0.71	1	4.10	0.76	2
Public housing	4.37	0.76	2	4.53	0.75	1
Garbage avoidance	4.24	0.73	3	3.82	0.77	4
Garbage disposal	3.82	0.84	4	3.60	0.86	9
Urban development/renewal	3.70	0.83	5	3.42	0.85	12
Social services	3.57	0.69	6	3.40	0.88	13
Youth welfare	3.57	0.68	7	3.22	0.82	16
Elementary schools	3.55	0.66	8	3.50	0.74	10
Higher education	3.54	0.70	9	3.46	0.70	11
Local economic development	3.51	0.90	10	3.20	0.83	17
Public health	3.50	0.71	11	3.64	0.75	8
Sewage disposal	3.48	0.73	12	3.36	0.78	15
Science	3.45	0.75	13	3.03	0.80	20
Culture	3.38	1.06	14	3.05	0.97	19
Parks and recreation areas	3.38	0.76	15	3.66	0.79	7
Energy supply	3.29	0.74	16	3.18	0.51	18
General welfare	3.23	0.77	17	2.78	0.92	21
Sport and leisure facilities	3.22	0.77	18	3.80	0.71	6
Public order and security	3.14	0.73	19	3.82	0.82	5
Parking lot construction	2.87	1.33	20	3.82	0.94	3
Street construction	2.66	1.11	21	3.39	0.94	14

QUESTION WORDING: * 'Please think about the following policy fields. What is your opinion on the structure of local expenditure in your city'? ** 'Please estimate the preference of the majority of voters in your city'. The response alternatives were: large cuts (1), some cuts (2), keep the same level of spending (3), some increase (4), large increase (5). N = 125–140.

shows, the link between the priorities of the elected and councilor perceptions of electors' opinions is stronger for core elements in respective policy profiles. Thus, CDU/CSU councilors do not deviate as strongly from their perceptions of public opinion on parking lot and street construction expenditure, nor on public order and security, as do Social Democrats, who see themselves as being less supportive of higher spending in these arenas than the electorate. A comparable picture is given for social services. This public service is given a lower priority by CDU/CSU councilors than they think the populace affords these services. At the same time, CDU/CSU politicians are more supportive of higher spending on the economic development, while public order and security is

Table 8.3 Spending Preferences of Politicians and Politicians' Assessments of Electoral Preferences, by Party of Politicians (Ranked on the Basis of Mean Values)

Item	CDU/CSU			SPD		
	Pop.	Pol.	Dif.	Pop.	Pol.	Dif.
Public transportation	2	1	+1	2	2	–
Public housing	1	2	–1	1	1	–
Garbage avoidance	5	3	+2	5	3	+2
Garbage disposal	9	5	+4	8	4	+4
Urban development/renewal	14	5	+9	11	8	+3
Social services	12	20	–8	15	5	+10
Youth welfare	15	17	–2	19	6	+13
Elementary schools	10	9	+1	10	10	–
Higher education	11	8	+3	13	11	+2
Local economic development	19	7	+12	16	131	+3
Public health	7	12	–5	7	9	–2
Sewage disposal	13	12	+1	14	7	+7
Science	18	11	+7	20	16	+4
Culture	20	15	+5	17	12	+5
Parks and recreation areas	6	14	–8	8	15	–7
Energy supply	17	16	+1	18	18	–
General welfare	21	21	–	21	14	+7
Sport and leisure facilities	4	19	–15	6	6	–10
Public order and security	8	10	–2	3	19	–16
Parking lot construction	3	4	–1	3	21	–18
Street construction	15	18	–3	11	20	–9

Pol. = Spending preferences of politicians. Pop = perceived spending preferences of the population. Dif. = difference of ranks between pop. and pol. Ranks are based on the average evaluation of CDU/CSU councilors on the one hand and SPD councilors on the other hand (with regard to the five point scale, corresponding to Table 8.2.) N-CDU/CSU = 45, N=SPD = 64.

ranked much lower on the Social Democrat agenda. Comparable results can be observed with respect to ideological orientations (for details see Gabriel et al., 1993).

As the results we have to hand clearly indicate, local issues and decisions are ranked and priorities set according to ideological commitments. The traditional view that local government is – or at least should be – nonpolitical has to be rejected. Left-right orientation and party affiliation have a strong impact on councilors' issue preferences. Only a weak positive relationship exists between the spending preferences of policy-makers and how politicians perceive electoral opinion. The weak linkage between the elected and the electorate is filtered by local politicians' ideological dispositions (Table 8.3).

THE IMPACT OF PARTIES: MACRO-LEVEL ANALYSES

Hypotheses

Irrespective of the view that local government is nonpolitical, the local party platforms of the CDU/CSU, SPD and FDP in the 1970s showed that the main political parties have long taken local self-government to be a serious 'arena of party politics' (Wollmann 1991, p.17).[7] Yet parties who neglect the preferences of their clientele have to expect exit and voice from the electorate (Hirschman, 1970). For this reason, Downs (1957, pp. 103ff., 109ff.) refers to the necessary 'integrity', 'reliability' and 'responsibility' of party platforms and, as a result, their 'coherence' and relative 'immobility'. Otherwise, the credibility of parties could suffer and, under the usual conditions of uncertainty, this could be a decisive determinant of voting decisions. Especially in multi-party systems, it might therefore be adequate for political parties to orient themselves toward ideologies instead of the expected demands of the majority of the electorate (Riker & Ordeshook, 1973, p. 338ff.). Accordingly, Downs assumes clear policy-differences between rightist and leftist parties in multi-party systems (under conditions of electoral uncertainty). These differences show up particularly strongly in attitudes toward interventions by the state in the economic processes (Downs, 1957, p. 112). Hence, an active public sector role in economic and social affairs should be typical of political agencies controlled by socialist or social democratic parties. On the other hand, rightist party dominance should be a factor leading to a restricted role of government in economic and social affairs.

The last 20 years, with its economic troubles, rising unemployment and intensifying fiscal crises, should have increased the politicization of the local arena. Political actors face increasing pressure to act, because of rising social problems and an enduring shortage of resources. Municipalities in Germany are important public employers, as well as the main investors of public funds. Within the limits of their financial possibilities, through their tax rates, they can work toward an anti-cyclical investment policy, with high local multiplier-effects promoting economic growth. But German cities are varied in their policies. Some invest much more than others, some employ many more people and some have much higher citizen and business taxes than others (Kunz, 1998, 2000). In line with the findings already presented, direct and intense partisan impacts on local policy differences might be expected. Cities where the Social Democrats hold a majority should be more active than others in social and economic affairs. This should lead to more municipal employees, higher personnel expenditure and higher tax rates. Accordingly, cities where the

CDU/CSU holds a majority should have comparatively fewer staff, lower spending on personnel and lower tax rates. Only in the field of public investment might knowledge of the party in power not enable one to predict expenditure levels. This is because, on the one side, almost any investment can be interpreted as contributing to employment, while, on the other side, local public investment can improve conditions for private economic activities (i.e. the neoliberal view of a temporary replacement of private activities through public investment). Hence, no substantial differences in investment expenditure between Social Democrat and Christian Democrat cities should be observed.

Operationalization and Research Design

Indicators for local employment, spending and revenue policies are dependent variables in this analysis. The policies examined are investment and personnel expenditure (per capita), municipal employment (per 10,000 residents) and local tax policies.[8] Tax policy is restricted to the business tax rate and the property tax rate B (*Hebesatz*). The property tax A is for agricultural property (including private woodland and horticulture) and is of minor significance for county-free cities. The business tax rate has a key impact on local revenue (for details on local taxation see Bennett & Krebs, 1988). The data on investment expenditure do not include grants from the state as these transfers are an explanatory variable in the analyses (Morss, 1966). The data on municipal employees refer to full-time employees in the municipal bureaucracy (without the four cities in the *Land* Schleswig-Holstein as data are not available for these cities).

The analyses encompass the period 1980 to 1989. Data on municipal finance before 1980 are not comparable with those after 1980, with 1989 being the last year for which reliable data are available (for more detail, see Kunz, 2000, p. 111ff.). Regarding the limited research on the causes of local fiscal decisions in Germany, this time limitation does not detract from the value of this study. The studies that exist mostly cover a few fiscal years, and often only one year (Pross, 1982; Deubel, 1984; Bennett, 1985; Hotz, 1987; Gruener et al., 1988; Bothe, 1989; Kunz & Zapf-Schramm, 1989; Eckey, 1991; Parsche & Steinherr, 1995; Pohlan, 1997). As municipal budget totals can vary 'by chance' from year to year, herein I analyze data for the whole of the 1980s.

To test the impact of political party control on policy commitments, four measures of party control are used.[9] With the city council as a quasi-legislative body similar to the national parliament, the first three capture the percentage of SPD and CDU/CSU members of the city council, plus the time between 1976 and 1989 that the Christian Democrats held a relative majority on the council. The fourth indicator is based on the length of time that the SPD has held the post of the mayor (*Oberbuergermeister*). With this political control measure the special strength of the mayor in the South German *Laender* during the 1980–1989 period is taken into account.[10] The characteristics contributing to the critical role of the mayoral position are numerous. The mayor is directly elected by the citizens, and not by the city council, for a relatively long term (8–10 years compared to 4–5 years for councilors). S/he is president of the city council and of all relevant council committees, and is in charge of preparing and chairing meetings of the city council, as well as executing decisions made by the council. Moreover, s/he has a formal vote in council decisions. At the same time, the mayor acts as the chief executive of the city administration and may give binding instructions to the heads of all administrative departments (for details, see Gunlicks, 1986, p. 73ff.; Gabriel et al., 1999, p. 338ff.).[11] Figure 8.1 shows the measurement model of party control.[12] A confirmatory factor and principal components analysis indicates that the four party control variables overlap in substantial ways. The different signs of the component loadings show that the principal dimension identified by the principal components analysis captures a left-right continuum. In the following analyses, scores on the first component are used as the index of party dominance. High component values refer to SPD-dominance, with low component values indicating CDU/CSU dominance.

Following the findings of a large number of international output studies, other variables were included as global indicators of the local structure of service needs and fiscal resources. As discussed in detail elsewhere (Kunz, 1998, 2000, pp. 164ff., 330ff.), the most powerful socio-economic explanatory variables are city urbanization, urban centrality, local social problems and investment grants by the state. Urbanization tends to generate greater demands for public services because of negative externalities. "Such negative externalities may include various forms of environmental pollution, traffic congestion and traffic accidents, and will call for policy measures within the area of environmental health, traffic regulation etc" (Hansen, 1984, p. 347). Urban centrality refers to the idea that cities provide services and facilities not only for their own population but for hinterland populations (Brainard & Dolbear, 1967; Aiken & Depré, 1981; Sharpe & Newton, 1984, p. 116ff.). These variables also directly affect the city tax base, because the density of economic activities and the value of commercial property, as well as the wealth of the commercial and business population, is higher in urbanized zones and in central cities (Postlep, 1985, p. 18ff.). To this extent, it might be hypothesized that the more urbanized the city and the higher the city rank in the central place hierarchy the more likely it is to spend on personnel and investment categories. Because of the concentration of economic activities, particularly in highly urbanized communities, these cities have important locational

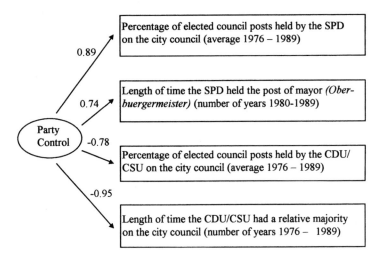

Comparative Fit-Index = 1.0, Bentler Bonett Normed Fit-Index = 1.0, Average Absolute Standardized Residuals-Index = 0.007 (Maximum-Likelihood-Solution). N = 87.

Figure 8.1. Measurement Model of Party Control (Standardized Factor Loading)

attractions (Hoover, 1948; Goldstein & Gronberg, 1984). Therefore, the more urbanized a city the less likely is the realization of exit threats from economic actors under the conditions of high taxation. Hence, urbanization should have a strong influence on the business tax rate, although a high tax base in urbanized cities might reduce this effect.

Social problems affect local services, financial resources and tax rates in different ways. Cities with many social problems have to provide a wide range of specialized social services, yet unemployment in particular has a negative impact on city fiscal resources. Because of this need-resource imbalance, the effect of social problems on tax rates should be positive, while the net effect on personnel and investment expenditure and municipal employment should be negative.

Investment grants by the state are the most important and most controversial factor determining local investment activities (Wilde, 1968; Boyne, 1990; Gabriel et al., 1990). Depending on one's interests, the influence of investment grants is criticized as a reduction of local autonomy or justified as contributing to a balanced spatial distribution of public infrastructure. What can be said is that, in times of scarce resources, communities lament the decline of investment grants by the state. Empirically, however, it remains to be shown that investment grants affect the investment spending of county-free cities.

Results

Table 8.4 reveals the results of the multivariate regression analyses for each of the policy measures investigated. Focusing on standardized beta weights, party control is one of the main predictors to have impacted on local policy-making.[13] Comparison of the direction of the party impact measure shows that differences between political parties in local policy-making primarily concern investment and consumption expenditure. CDU/CSU influence is correlated with high levels of investment expenditure. On the other hand, SPD influence correlates with more municipal employees and higher personnel expenditure. This shows that parties in local politics – at least in the cities analyzed – realize different political objectives. In the fields of personnel expenditure and municipal employment this is to be expected, for the party principles of the Social Democrats are directed at demand-oriented compensation for socio-economic needs. More surprising is the correlation between CDU/CSU dominance and higher investment expenditure. Yet it is true that a sizable segment of local public investment spending includes commitments that assist private sector activities, so the direction of CDU/CSU impact is compatible with party principles. Significantly, with the SPD associated with high spending in one policy arena, and the CDU/CSU being associated with higher commitments in another field, a weak ideological impact is found for party impacts on tax rates. That said, the effect on the business rate is far from being a quantité négligeable.

As Table 8.4 shows, socio-economic variables are rarely stronger predictors than party control. The place of a city in the central place hierarchy does have strong effects on investment and personnel expenditure, as well as municipal employment, which confirms the hypothesis that greater urban centrality generates more demand for public services. Measuring demand for local services solely on the characteristics of those living in a city therefore underestimates actual demand. But low and even negative effects on tax rates suggests that urban centrality also improves municipal fiscal prospects. In terms of placing these conclusions on urban centrality alongside those from previous investigations, for personnel expenditure the results correspond with previous findings from Germany. For investment expenditure, by contrast, no clear results emerge from previous work (Schwarting, 1979; Reissert, 1984; Gabriel et al., 1990). One reason is that previous research has faced substantial problems over measuring urban centrality, as the dimensions urbanization and centrality could not be separated statistically. For the cities in this study, urbanization does not

Table 8.4 Determinants of Municipal Policy Outputs (Standardized Beta Weights in Regression Analysis)

Explanatory variable	Dependent variable (average 1980 – 1989)				
	Investment expenditure (per head)	Personnel expenditure (per head)	Municipal employees (per 10,000 residents)	Business tax rate (*Hebesatz*)	Property tax rate (*Hebesatz* property tax B)
Party dominance (SPD vs. CDU/CSU)	–0.32	0.35	0.29	0.16	0.11
Urban centrality	0.40	0.36	0.28	–0.07	–0.12
Urbanization	0.10	0.35	0.28	0.44	–0.09
Social problems	–0.35	0.00	0.06	0.07	0.19
Investment grants	0.17				

City centrality was assessed by a factor index using labor market centrality (employment commuters into the city minus employment commuters out of the city, divided by the employed city population, in 1987), educational centrality (those commuting for training into the city minus those commuting out of the city for training, divided by the city population in training, in 1987) and market service centrality (employees in market services per 1,000 residents, averaged for 1982–1989). The factor loadings for these three measures were 0.89, 0.86 and 0.81. Urbanization was represented by a factor index, using population size (the average for 1980–1989), population density (residents per square kilometer of the city area, averaged for 1980–1989), settlement density (residents per square kilometer of the settlement area, averaged for 1980–1989) and employee density (employees per square kilometer of the city area, averaged for 1980–1989). The factor loadings for these variables were 0.88, 0.87, 0.84 and 0.80. Social problems were represented by the rate of unemployment (averaged for 1985–1989) and the percentage of people receiving welfare (averaged for 1980–1989). The index for this measured was comprised of the factor loadings for these variables, which added and divided by two. Investment grants were represented by receipts from the state per head (averaged over 1980–1989). Other variables that were included, but had only negligible effects, were population change, age structure of the population and share of employees in technical and consultancy services. Sources: *Statistisches Jahrbuch Deutscher Gemeinden, Bundesforschungsanstalt fuer Landeskunde und Raumordnung (Bundesamt fuer Raumwesen und Raumordnung)*. The correlations of the explanatory variables are all between |0.07| and |0.49| and the condition number of the correlation matrix is far below 100. Hence, multicollinearity problems are not of importance (Farrar & Glauber, 1967; Belsley et al. 1980, p. 153).

correlate with position in the central place hierarchy (r = 0.01). The fact that urban centrality is related to differences in personnel expenditure (Table 8.4), but not variation in investment spending, might be a product of above-average infrastructural development in urban regions (Kunz 1991).

Not only do municipal employment and personnel expenditure increase alongside levels of urbanization, but so to do business tax rates. The relationship with business taxes is much stronger than that recorded for any other socioeconomic or party variable. Possibly this might be a result of special locational factors in urban communities (e.g. Hoover, 1948; Goldstein & Gronberg, 1984), which means that businesses find more urban locations sufficiently attractive to outweigh the disadvantages higher taxes, which enables local councils to levy higher rates with relative impunity (for a recent analysis of the attractions of city location for modern businesses, see Bennett et al., 1999).

Social problems have a negative impact on investment expenditure but a weak positive effect on the property tax rate. In this regard, multivariate quantitative analysis confirms the often expressed opinion that local investment spending depends substantially on fiscal resources, while rising social problems limit resources, so cuts in investment expenditure are inevitable. The weak to negligible impact of social problems on tax rates shows that in the search to procure additional resources city governments in Germany seek to spare tax payers. In line with Peterson's (1981) suggestions for U.S. cities, it appears that the leaders of cities with substantial social problems seek to respond to these problems in a manner that does not impact in major ways on tax rates, in order to maintain or to promote a favorable climate for business and higher income residential investment.

The effect of investment grants by the state on local investment expenditure is moderate. This result might be partly a product of a drastic decline in investment grants in the 1980s. But the more plausible explanation, which Reissert (1984) suggests in his extensive analysis of local investment expenditure from 1956 to 1970, is that the link between investment grants and investment expenditure is weak. The results of this study certainly seem to point in this direction. It is clear that total investment expenditure is determined by the importance of a city in the central place hierarchy, by the magnitude of local social problems and by political party control, with the effects of these variables persisting when national government investment grants are taken into account.

CONCLUSION

Contrary to the legal-institutional approach in German research on local government, this chapter offers an empirical investigation of political party impacts on local government policy. Political parties have a decisive role in the shaping of local policy decisions in Germany. This applies to the selection of political leaders, including local council members, as well as to the choice of public policies. Irrespective of strong reservations made by German theorists of the constitution regarding the appropriate role of political parties in local self-administration, the results of micro- and macro-level analyses show the crucial role of parties in local politics. Party ideology weakens the relationship between the spending preferences of local policy-makers and their assessment of the electorate's preferences. In addition, as multivariate analysis demonstrates, political party control impacts on both municipal budgets and employment. Communities governed by the SPD have expanded the size of the municipal bureaucracy, whereas municipalities in which the CDU/CSU is the leading party have favored capital investment that brings potential private sector benefits. Socio-economic variables are rarely distinguished as stronger predictors of municipal budgets and employment than party control.[14]

NOTES

1. Apart from data problems (e.g. data on the Green Party are only available since 1984), the consideration of further cleavages would lead to groups with a very small number of councilors and/or city-councils under their control, so party effects for them can not be measured reliably.

2. The data were collected by a mail questionnaire sent to all 338 councilors in the county-free cities Bamberg, Bonn, Heilbronn, Luwigshafen, Ulm and Wiesbaden. The survey in Bamberg, Bonn, Heilbronn and Luwigshafen was conducted in 1991. Surveys in Ulm and Wiesbaden were conducted in 1993. The response rate was 59% (project management by Oscar W. Gabriel). Correlations were computed controlling for the city of each councilor and for the date of survey (as dummy variables), but the impact of these measures was meager, so these results are not presented here. Significance levels are not presented as the survey did not use a random sample. The impact of subjective dispositions on local fiscal policies cannot be analyzed because of the small number of cities in this data set.

3. Respondents were asked to place themselves on an eleven-point left-right scale.

4. Public security is a local administrative responsibility with heavy personal contributions. Despite this, the correlations show that strong law-and-order orientations are not associated with the favoring of higher spending on municipal employment.

5. These variables were only collected in the survey in 1991 in the cities of Bamberg, Bonn, Heilbronn and Ludwigshafen.

6. The question of whether perceived preferences of the population correspond to actual citizen preferences is not answered here. To address this, surveys of the electorate in each city would be necessary.

7. In the 1980s, these platforms were partially renewed and extended. In 1987, the FDP published *Guidelines of Liberal Local Politics.* In 1989, the CSU published a special program on local politics (*Safe Future in a Home Worth Living*; for details see Knemeyer and Jahndel 1991, pp.47ff., 34ff.). For differences between local councils and national government, see Gabriel (1984).

8. Sources: *Statistiches Jahrbuch Deutscher Gemeinden, Statistische Landesaemter, Bundesforschungsantalt fuer Landeskunde und Raumordnung (Bundsamt fuer Raumwesen und Raumordnung).*

9. Sources: *Statistisches Jahrbuch Deutscher Gemeinden, Sozialdemokratische Gemeinschaft fuer Kommunalpolitik.*

10. Until 1990 four different municipal charters existed in the German Laender: the South German Council form in Baden-Wuerttemberg and Bavaria, the North German counterpart in North-Rhine-Westphalia and Lower Saxony, the Magistrate form in Hesse

and cities in Schleswig Holstein, and the Strong Mayor form in Rhineland-Palatinate, Saarland and rural communities in Schlewsig-Holstein. These four institutional arrangements may be placed along a continuum regarding the strength of the two leading institutions in the local political system – the city council and the mayor. Due to various institutional regulations, a clearly dominant position is attributed to the mayor in the South German council form.

11. During the last decade, several local government reforms have been implemented in Germany involving crucial elements of municipal charters. According to their promoters, these reforms had two main objectives: to increase the efficiency of local political institutions and to give people more say in local political life. Today, all mayors are elected by popular vote.

12. Further measures of party control, such as the length of time the Social Democrats held a relative majority on the city council or the difference between the percentages of elected representatives of the SPD and CDU/CSU, could not be included because they were highly correlated with other political variables.

13. The level of significance is not relevant here, as the sample of cities is not random.

14. I am grateful to Keith Hoggart and Catrin Yazdani for revising the initial draft of this manuscript.

REFERENCES

Aiken, M., & Depré, R. (1981). The Urban System, Politics, and Policy in Belgian Cities. In: K. Newton (Ed.), *Urban Political Economy*, (85–116). London: Frances Printer.

Alcaly, R. E., & Mermelstein, D. (1977). *The Fiscal Crises of American Cities*. New York: Vintage Books.

Belsley, D. A., Kuh, E., & Welsch, R. E. (1980). *Regression Diagnostics: Identifying Influential Data and Sources of Collinearity*. New York: Wiley

Bennett, R. J. (1985). The Finance of Cities in West-Germany. *Progress in Planning* 21, 1–62.

Bennett, R. J., & Krebs, G. (Eds.). (1988). *Local Business Taxes in Britain and Germany*. Baden-Baden: Nomos.

Bennett, R. J., Graham, D. J., & Bratton, W. (1999). The Location and Concentration of Businesses in Britain: Business Clusters, Business Services, Market Coverage and Local Economic Development. *Transactions of the Institute of British Geographers* 24, 393–420.

Bergstrom, T. C., & Goodman, R. P. (1973). Private Demand for Public Goods. *American Economic Review 63*, 280–296.

Bothe, A. (1989). *Die Gemeindeausgaben in der Bundesrepublik: Ein Nachfrageorientierter Erklaerungsansatz*. Tuebingen: Mohr.

Boyne, G. A. (1990). Central Grants and Local Policy Variation. *Public Administration* 68, 207–233.

Boyne, G. A. (1992). Local Government Structure and Performance: Lessons from America?. *Public Administration 70*, 333–357.

Brainard, W. C., & Dolbear, F. T. (1967). The Possibilities of Oversupply of Local Public Goods: A Critical Note. *Journal of Political Economy* 75, 86–90.

Dahl, R. A. (1971). *Polyarchy. Participation and Opposition*. New Haven: Yale University Press.

Deubel, I. (1984). *Der Kommunale Finanzausgleich in Nordrhein-Westfalen: Eine Oekonomische und Statistische Analyse*. Koeln: Deutscher Gemeindeverlag Kohlhammer.

Downs, A. (1957). *An Economic Theory of Democracy*. New York: Harper and Row.

Eckey, H-F. (1991). Ermittlung des Finanzbedarfs im Hessischen Kommunalen Finanzausgleich - unter Besonderer Beruecksichtigung Statistischer Verfahren. In: Innenminster des Landes Nordrhein-Westfalen (Ed.). *Die Bedarfsermittlung im Kommunalen Finanzausgleich: Ergebnisse eines Erfahrungsaustausches von Praktikern und Wissenschaftlern*, (107–122). Berlin: Analytica.

Farrar, D. E., & Glauber, R. R. (1967). Multicollinearity in Regression Analysis: The Problem Revisited. *Review of Economic and Statistics 49*, 92–107.

Frank, R. (1990). *Kultur auf dem Pruefstand: Ein Streifzug durch 40 Jahre Kommunaler Kulturpolitik*. Muenchen: Minerva.

Fried, R. C. (1976). Party and Policy in West-German Cities. *American Political Science Review 70*, 11–24.

Gabriel, O. W. (1984). Parlamentarisierung der Kommunalpolitik. In: O.W. Gabriel, P. Haungs & M. Zender (Eds.), *Opposition in Großstadtparlamenten*, (101–147). Melle: Verlag Ernst Knoth.

Gabriel, O. W. (2000). Democracy in Big Cities: The Case of Germany. In: O. W. Gabriel, V. Hoffmann-Martinot & H. V. Savitch (Eds.), *Urban Democracy*, (187–260). Opladen: Leske and Budrich.

Gabriel, O. W., Brettschneider, F., & Kunz, V. (1993). Responsivitaet Bundesdeutscher Kommunalpolitiker. *Politische Vierteljahresschrrift 34*, 29–46.

Gabriel, O. W., Kunz, V., & Zapf-Schramm, T. (1990). *Bestimmungsfaktoren des Kommunalen Investitionsverhaltens*. Muenchen: Minerva.

Gabriel, O. W., Kunz, V., & Ahlstich, K. (1999). Die Kommunale Selbstverwaltung. In: O. W. Gabriel, & E. Holtmann (Eds.), *Handbuch Politisches System der Bundesrepublik Deutschland*, (325–354). second edition. Muenchen: Oldenburg.

Gau, D. (1990). *Kultur als Politik: Eine Analyse der Entscheidungspraemissen und des Entscheidungsverhaltens in der Kommunalen Kulturpolitik*. Muenchen: Minerva

Goldstein, G .S., & Gronberg, T. J. (1984). Economies of Scope and Economies of Agglomeration. *Journal of Urban Economies 16*, 91–104.

Gruener, H., Jaedicke, W., & Ruhland, K. (1988). Rote Politik im Schwarzen Rathaus?: Bestimmungsfaktoren der Wohnungspolitischen Ausgaben Bundesdeutscher Großstaedte. *Politische Vierteljahresschrift 29*, 42–57.

Gunlicks, A. (1986). *Local Government in the German Federal System*. Durham, North Carolina: Duke University Press.

Haeussermann, H., & Siebel, W. (1986). Die Polarisierung der Großstadtentwicklung im Sued-Nord-Gefaelle. In: J. Friedrichs, H. Haeussermann, & W. Siebel (Eds.), *Sued-Nord-Gefaelle in der Bundesrepublik?: Sozialwissenschaftliche Analysen*, (70–96). Opladen: Westdeutscher Verlag.

Haeussermann, H., Petrowsky, W., & Pohlan, J.. (1995). Entwicklung der Staedte: Stabile Polarisierung. In: T. Jaeger & D. Hoffmann (Eds.), *Demokratie in der Krise?: Zukunft in der Demokratie*, (191–213). Opladen: Leske and Budrich.

Hansen, T. (1984). Urban Hierarchies and Municipal Finances. *European Journal of Political Research 12*, 343–356.

Heinelt, H. 1991. Die Beschaeftigungskrise und arbeitsmarkt- und sozialpolitische Aktivitaeten in den Staedten. In: H. Heinelt, & H. Wollmann (Eds.), *Brennpunkt Stadt: Stadtpolitik und Lokale Politikforschung in den 80er und 90er Jahren*, (257–280). Basel: Birkhaeuser.

Hirschman, A. O. (1970). *Exit, Voice and Loyalty*. Cambridge, Massachusetts: Harvard University Press.

Hoggart, K. (1989). *Economy, Polity and Urban Public Expenditure*. Aldershot: Avebury.

Holtmann, E. (1992). Politisierung der Kommunalpolitik und Wandlungen des Lokalen Parteiensystems. *Aus Politik und Zeitgeschichte* B 22–23/1992, 13–22.
Hoover, E. M. (1948). *The Location of Economic Activity*. New York: McGraw-Hill.
Hotz, D. (1987). Arbeitslosigkeit, Sozialhilfeausgaben und Kommunales Investitionsverhalten. *Informationen zur Raumentwicklung* 1987, 593–610.
Jaedicke, W., Ruhland, K., Wachendorfer, U., & Wollmann, H. (1990). *Lokale Politik im Wohlfahrtsstaat: Zur Sozialpolitik der Gemeinden und ihrer Verbaende in der Beschaeftigungskrise*. Opladen: Westdeutscher Verlag.
Klingemann, H-D., & Volkens, A. (1997). Struktur und Entwicklung von Wahlprogrammen in der Bundesrepublik Deutschland 1949–1994. In: O. W. Gabriel, O. Niedermayer & R. Stoess (Eds.), *Parteiendemokratie in Deutschland*, (517–536). Bonn: Bundeszentrale fuer Politische Bildung.
Klingemann, H-D., Hofferbert, R. I., & Budge, I. (1994). *Parties, Policies, and Democracy*. Boulder: Westview Press.
Knemeyer, F-L., & Jahndel, K. (1991). *Parteien in der Kommunalen Selbstverwaltung*. Stuttgart: Boorberg.
Kunz, V. (1991). Infrastruktur, Betriebsgröße und Höherwertige Tertiärisierung als Bestimmungsfaktoren der Regionalen Wirtschaftskraft. *Informationen zur Raumentwicklung* 1991, 579–598.
Kunz, V. (1998). Die Hebesatzpolitik der Kreisfreien Staedte in den 80er Jahren. In: H. Maeding & R. Voigt (Eds.), *Kommunalfinanzen im Umbruch*, (161–184). Opladen: Leske and Budrich,.
Kunz, V. (2000). *Parteien und Kommunale Haushaltspolitik im Staedtevergleich*. Opladen: Leske and Budrich.
Kunz, V., & Gabriel, O. W. (1992). Determinanten der Kommunalen Kulturausgaben. *Informationen zur Raumentwicklung* 1992, 43–55.
Kunz, V., & Zapf-Schramm, T. (1989). Ergebnisse der Haushaltsentscheidungsprozesse in den Kreisfreien Staedten der Bundesrepublik. In: D. Schimanke (Ed.), *Stadtdirektor und Buergermeister: Beitraege zu einer Aktuellen Kontroverse*, (161–189). Basel: Birkhaeuser.
Lehmbruch, G. (1979). Der Januskopf der Ortsparteien. Kommunalpolitik und das Lokale Parteiensystem. In: H. Koeser (Ed.), *Der Buerger in der Gemeinde*, (320–334). Bonn: Dietz, 320–334.
Lijphart, A. (1984). *Democracies: Patterns of Majoritarian and Consensus Government in Twenty Two Countries*. New Haven: Yale University Press.
McGovern, K. (1997). *Wirtschaftsfoerderung und Kommunalpolitik: Koordination und Kooperation*. Opladen: Leske and Budrich.
Mielke, B. (1985). Interkommunale Ausgabenunterschiede und Strukturmerkmale von Gemeinden am Beispiel Nordrhein-Westfalen. In: Raeumliche Aspekte des Kommunalen Finanzausgleichs (Ed.). *Akademie fuer Raumforschung und Landesplanung*, (99–131). Hannover: Vincentz.
Miera, S. (1994). *Kommunales Finanzsystem und Bevoelkerungsentwicklung*. Frankfurt/M.: Lang.
Morss, E. R. (1966). Some Thoughts on the Determinants of State and Local Expenditures. *National Tax Journal* 19, 95–103.
Pammer, W. J. (1990). *Managing Fiscal Strain in Major American Cities: Understanding Retrenchment in the Public Sector*. New York: Greenwood.
Parsche, R., & Steinherr, M. (1995). *Der Kommunale Finanzausgleich des Landes Nordrhein-Westfalen*. Muenchen: Ifo Institut fuer Wirtschaftsforschung.
Peterson, P. E. (1981). *City Limits*. Chicago: University of Chicago Press.

Pohlan, J. (1997). *Finanzen der Staedte: Eine Analyse der Mittelfristigen Entwicklungsunterschiede.* Berlin: Analytika.

Postlep, R.-D. (1985). *Wirtschaftsstruktur und Großstädtische Finanzen: Einflüsse höherwertiger Dienstleistungen auf die Kommunalen Steuereinnahmen und -Ausgaben in Verschiedenen Großstädten.* Hannover: Vincentz.

Pross, G. (1982). *Die Nachfrage nach Oeffentlichen Guetern: Eine Empirische Untersuchung der Gemeindeausgaben Baden-Wuerttemberg.* Frankfurt/M.: Fischer.

Puettner, G., (Ed.). (1982). *Handbuch der Kommunalen Wissenschaft und Praxis.* second edition. Berlin: Springer.

Reissert, B. (1984). *Staatliche Finanzzuweisungen und kommunale Investitionspolitik, Dissertationsschrift.* Berlin: Freie Universität Berlin.

Reuter, L-R. (1976). Kommunalpolitik im Parteienvergleich. Zum Funktionswandel der Kommunalen Selbstverwaltung anhand der Kommunalpolitischen Grundsatzprogramme von CDU/ CSU, SPD und FDP. *Aus Politik und Zeitgeschichte* B 34/1976: 3–45.

Rickards, R. C. (1985). Ursachen fuer die Nichtinkrementale Bildung von Haushaltsprioritaeten in Bundesdeutschen Staedten. *Archiv fuer Kommunalwissenschaften 24,* 295–309.

Riker, W. H., & Ordeshook, P. C. (1973). *An Introduction to Positive Political Theory.* Englewood Cliffs, New Jersey: Prentice-Hall.

Schmidt, M. G. (1996). When Parties Matter: A Review of the Possibilities and Limits of Partisan Influence on Public Policy. *European Journal of Political Science 30,* 155–183.

Schwarting, G. (1979). *Kommunale Investitionen: Theoretische und Empirische Untersuchungen der Bestimmungsgründe Kommunaler Investitionstätigkeit in NRW 1965–1972.* Frankfurt: Peter Lang.

Sharpe, L. J., & Newton, K. (1984). *Does Politics Matter?: The Determinants of Public Policy.* Oxford: Clarendon.

Wilde, J. A. (1968). The Expenditure Effects of Grant-In-Aid Programs. *National Tax Journal* 21, 340–348.

Winkler-Haupt, U. (1988). *Gemeindeordnung und Politikfolgen: Eine Vergleichende Untersuchung in vier Mittelstaedten.* Muenchen: Minerva.

Wollmann, H. (1991). Lokale Politikforschung und Politisch-Gesellschaftlicher Kontext: Eine Entwicklungsskizze am Beispiel des Arbeitskreises Lokale Politikforschung. In: H. Heinelt & H. Wollmann (Eds.), *Brennpunkt Stadt: Stadtpolitik und Lokale Politikforschung in den 80er und 90er Jahren,* (15–30). Basel: Birkhaeuser.

Wright, W. E. (1971). Party Models: Rational Efficient and Party Democracy. In: W. E. Wright (Ed.), *A Comparative Study of Party Organizations,* (85–114). New York: Basic Books.

9. LOCAL GOVERNMENT GROWTH AND RETRENCHMENT IN PORTUGAL: POLITICIZATION, NEO-LIBERALISM AND NEW FORMS OF GOVERNANCE[1]

Carlos Nunes Silva

In the late 1970s and especially during the 1980s and 1990s, local government in most industrialized countries was confronted with fiscal difficulties. Local institutions were in the middle of a political storm centered on downsizing the state. Conservative and neo-liberal policies made headway, with a focus on rolling back the state, privatization and reducing public expenditure. For Portugal these decades saw a shift from municipalities being key local agents to a situation in which decision-making and service provision is increasingly shared by public and private entities. In the process, new forms of urban governance have been created, leading to more complex institutional environments. The traditional form of local public administration has given way to public-private partnerships. Local government spends more, but is no longer involved or influential in many arenas of social and economic importance. But what are the real effects of neo-liberal ideology on Portuguese local government? What is the real meaning of the new urban governance? These questions are at the heart of this chapter.

LOCAL GOVERNMENT POLICY CHANGE: MODES OF EXPLANATION

Interpretations of public policy shifts, whether involving growth or retrenchment, face several theoretical and methodological difficulties (Duncan & Goodwin, 1988; Walsh, 1988) and the literature on the topic is inconclusive (Tarschys, 1975; Silva, 1995a). Research on government growth identifies numerous theories, most of which have constraint or choice as the primary focus of attention (Newton, 1974; Miranda & Walzer, 1994). This uncertainty existed before the emergence of new forms of governance, which has complicated debate further (e.g. Judge et al., 1995; Pierre, 1998).

The changing environment in which local government acts has undoubtedly complicated the task of explaining local government policy determination. For one, there has been a growing dependence of local political institutions on the involvement of civil society in policy making and implementation, even if such involvement in urban government has in some measure occurred over most of the twentieth century. What seems to be new is the extent of collaboration in almost all fields of public policy. This change is particularly important at the local level: (a) because local political institutions are less complex than national ones and are therefore more open to civil society; (b) because they have more limited resources, so there is greater interest in obtaining additional inputs from civil society; (c) because they are less professionalized, with the consequence that links between local politicians and civil society are closer, with more opportunities for interaction; and, (d) because local policy processes have a higher degree of integration of strategies of different actors, so there are more opportunities for public-private co-operation than at the national level. A second change is related to the development of welfare-states, which has been accompanied by pressures for more efficient and effective public service production and delivery. As a consequence, managerial problems have become a central issue in local government in most states in Europe (and elsewhere). Fiscal problems in the 1970s and 1980s increased interest in managerial issues in countries with very different levels of economic development, with the consequence that ideas on 'new public management' and consumer choice gained more currency, just as market-oriented approaches to service delivery gained increased importance (Cochrane, 1993; Stewart & Stoker, 1995; Walsh, 1995; Keen & Scase, 1998).

Across Europe, local government is increasingly using these new instruments to enhance the capacity to act. As a consequence, *urban governance* is becoming more of a public-private affair, contrary to the old system of public administrative exclusiveness. As well as being driven by service considerations, an important factor in the development of the new urban governance is the revitalization of local democracy (Burns et al., 1994, p. 31). Yet the key question of the real consequences of new forms of urban governance for local government policy remain.

THE POLITICIZATION OF LOCAL GOVERNMENT IN PORTUGAL

In Portugal, pressures favoring new forms of urban governance are complicated by unprecedented levels of political change. The current local government system is based on the Constitution of 1976, which was adopted after the Revolution of 1974. Before then municipalities and parishes were two tiers of local administration. They were strictly controlled by the national government, with no elected members (Caetano, 1983; Amaral, 1993; Oliveira, 1993). Since the December 1976 elections in municipalities and parishes, the principle of representative local democracy has been a fundamental feature of the Portuguese political system. Yet local government is a relatively feeble arm of the state. This might not appear to be the case given the size of local government institutions, as they service quite large populations. Thus, for the Portuguese population of just short of 10 million people, there are 4,241 parishes (*freguesias*) and 308 municipalities. This gives an average of 32,500 municipal inhabitants,[2] with the parish figure at around 2,400. These units only account for about 10% of public expenditure. Moreover, most municipalities depend largely upon block grant receipts from the national government for their revenue. Indeed, the local finance system allows limited autonomy to municipalities.[3] In comparative terms, Portugal is one of the most centralized countries in the European Union (EU). Within the EU, Portugal is only ahead of Greece in terms of the ratio of municipal revenue to GDP or the share of municipal spending in total state expenditure (Silva, 1998). Although local government has a general power of local competence, in reality the most important local public services are associated with urban planning, basic infrastructure provision and maintenance services. Social welfare and economic development are secondary functions compared to national actions in these fields (Sá, 1991; Silva, 1995). Adding to the rather 'low key' feel of local government, few elected councilors work full-time. The largest cities of Lisbon and Porto are allowed four full-time representatives, other municipalities with 100,000 or more inhabitants are allowed three, places with 20,000–100,000 have two, and smaller places can only have one full-time councilor. It requires a decision of the municipal executive council to exceed this number. Such approval is rarely granted, whether due to political party differences within councils themselves, or on account of worries about the impact of adding an extra salary to the municipal budget. Hardly surprisingly, it is not uncommon to hear criticism, especially from minority party members, without executive responsibility, that their participation in council affairs is limited to two meetings a month, with discussion and the necessity to make a vote having to be taken without full knowledge of their consequences. In most cases there is a sense of frustration and exclusion about the policy-making process. Added to which remuneration for full-time representatives is low, which partially explains the resistance of some professionals to accept full-time posts.

Arguably, compared with some other nations in Europe, the ability of Portuguese local governments to act as effective institutions is further complicated by the municipal electoral system. Under this system, two or more parties can get members elected onto the local executive board,[4] so it is not uncommon for no party to have an overall majority (e.g. Table 9.1). This frequently results in different parties forming a joint administration, under a formal or informal coalition, which usually involves the sharing of chairs (*pelouros*). But there are also minority administrations, in which one party forms an administration with the implicit support of another party. In sum, there is no uniform pattern of organization and operation for executive boards. Nevertheless, decisions on the budget, on fees, on the scale of services provided, on the municipal housing stock, or on leisure facilities, are all political decisions, irrespective of the party geometry on a municipal council.

Significantly in this regard, the local government system is experiencing an intensification of party politicization, as seen in the majority party automatically taking all posts of responsibility (*pelouros*). This is contrary to the practice of the 1970s,[5] with most decisions in the first years after 1974 taken on a consensual basis.[6] This has changed gradually since the end of the 1980s. Only a small number of councilors regard their party as important solely for elections and ignore it between elections. Most commonly, party groups vote as a bloc, both on the executive and on deliberative boards. For the latter, parties are generally organized as parliamentary units, with the party leadership directing assembly meetings. Parties are not simply mechanisms for securing election, but represent certain interests, even if there are differences in councilor accountability across local party organizations. Political recruitment also varies from party to party. Extensive party experience is not a necessary condition,[7] but, once selected as a candidate by a party, acceptance of party control is generally expected once elected. In reality, the local government system is based on a small group of political activists. Added to which there is unevenness in the role of each party's central office in local affairs, with central office attention stronger in the

Table 9.1 Number of Presidencies (Mayors) in Câmara Municipal (CM)

Election year	PSD	CDS/PP	AD	PS	PCP	right & centre-right	left & centre-left	Total
1976	115	36	–	115	37	152	152	304[(1)]
1979	101	20	73	60	50	195	110	305[(1)]
1982	88	27	49	84[(2)]	55	166	139	305[(1)(3)]
1985	149	27	–	79	47	176	129	305[(4)]
1989	113	20	–	116	50	134	171	305[(5)]
1993	116	13	–	126*	49	175	130	305[(6)]
1997	127	8	–	127*	41	135	169	305[(7)(8)]

Source: own computations from STAPE.
Notes:
(1) Includes one CM under PPM (Monarchy Party).
(2) Include one CM in coalition with UEDS.
(3) Include one CM under ASDI, in Azores.
(4) Include three CM under PRD.
(5) Include five coalitions (one CDS/PSD, three PS/CDS and one PS/PCP-MDP-PEV) and one CM controlled by UDP in Madeira.
(6) Includes one coalition in Lisbon (PS, PCP, PEV, PSR, UDP)
(7) One municipality controlled by PPM
(8) After this election three new municipalities were created (total = 308).
AD refers to acoalition between the PSD & the CDS.
PCP refers to candidates presented as a coalition of the FEPU, the APU, and the CDU.
* One more municipality in Lisbon could be added here which is controlled by a coalition.

major urban municipalities. In addition, there is uneven central control across parties, with the Communist Party having a stronger party input into councilors' activities (Silva, 1995, pp. 51–52).[8] But control is in no sense absolute, as there are cases in which the local party branch does not follow national directives (*Público*, 1994).[9]

Contrary to other European nations in which an administrative form of local politics is prevalent, in Portugal politicians rather than administrators dominate the policy-making process. For instance, in a 1994 questionnaire of mayors,[10] 71% held that councilors always initiated policies, with 35% holding that party politics was integral to day-to-day executive management. Before 1974, the *Secretário Municipal* (the Municipal Administrator) was more powerful than today's chief officers.[11] The relative influence of parties and officials has changed over time, as 'politics' has become more formalized. One aspect of this has been the provision of working rooms for all political parties who have elected members, with arrangements made for officers to advise even opposition parties. At the same time, the parties all produce detailed policy statements, which are translated into official council policy after elections (especially in the Annual

Municipal Activity Plan and the Annual Budget). Municipal departments are certainly expected to implement a party's election manifesto as policy.[12] In rural municipalities manifestos are usually a short document, with brief statements of intent, but they are generally more comprehensive in urban municipalities. Adding to party discipline, there is a widespread practice of party groups meeting before full council meetings in order to secure unity within the group over decisions that have to be taken by the council.

That said, there is evidence of a degree of factionalism within parties. Principally this is due to the influx of younger councilors and the withdrawal of founder leaders of the democratic regime. The 'distinction' this has created is between older members who place emphasis on service provision and younger councilors who are more concerned with value for money. In addition, there is undoubtedly some ideological passion in local government, with the left emphasizing the need for greater decentralization and egalitarian policies and the political right stressing efficiency and market criteria. On occasions, when key leadership changes have occurred, for instance when party control is lost after 20 years with an overall majority, relations between councilors can become distasteful, passionate and vituperative. These sentiments can spread beyond the walls of the council chamber, producing a situation that is far from the idea of a consensual local politics.[13] In this context it is worth noting that meetings involving verbal and physical abuse are not rare, whether these incidents involve councilors from different parties, councilors and protestors in the public galleries, or demonstrations held inside or outside council meeting rooms or buildings.

Despite such tensions, it is important to recognize the role of officers in influencing local councilors, even in a system where there is no powerful 'General Secretary' or 'Administrator' but simply directors of service departments (Silva, 1995, p. 61). For one, pressures from the general public are, in most cases, first mediated through the officer structure. In addition, there is increasing delegation of routine decision power to local officers, with proposals already made to introduce an administrator in the municipal apparatus who will alleviate the pressure of routine internal work on local councilors.[14] Some policy initiatives come from officers, in spite of the fact that 71% of mayors believed that they always take the initiative in the formulation of policy measures. What should nonetheless be noted is that increased public scrutiny and public participation requires the closer involvement of local councilors in decisions. At the same time, the trend toward contracting-out and 'privatizing' traditional municipal activities is eroding the influence of officers. The growing number of activity benchmarks emanating from the EU and the national government also imposes more standards on local officials, which further reduces their scope for independent action in many policy arenas (e.g. for environmental infrastructure, urban transportation, education and planning).

In spite of a growing political party presence, public involvement in local government affairs is low. Turnout in local elections is lower (60.1% in 1997) than in national elections (66.3% in 1995 and 61.9% in 1991). By way of comparison, in western nations turnout in city elections is on average around 70% (Rallings et al., 1996, p. 64). Perhaps more notable than this disparity are the conclusions of mayors in the 1994 questionnaire. Here 65% considered public participation in municipal meetings to be 'weak' or 'very weak'. Yet there seems to be growing involvement over environmental issues and over citizen rights (albeit in some cases at a rhetorical level), and there is a growing propensity for single-issue groups to challenge traditional local party politics. Moreover, alternative forms of democratic participation, such as local referenda and citizen-initiated ballots, although in the latter case lacking any legal status (i.e. they are merely indicative), are beginning to make more regular appearances.[15] Set against this, the tendency of municipalities to externalize the services they provide is weakening the centrality of local government as a focus of citizen interest, which is expected to dilute direct citizen political participation.

The reader should not come away with a picture of rather lethargic institutional behavior, where constraints on local government lead to policy change originating largely at the national level. In reality, there has been considerable change in service organization in recent years. Most evidently this has resulted from the reorganization of local government management, in which the search for improved efficiency and functional rationality has stimulated innovation. This has arisen not only as a consequence of new local councilors entering the system but also as a consequence of a general increase of technical staff qualifications. All major political parties accept the need for new instruments and forms of local government management as part of a more general approach to good practices in local government. The idea of residents as 'citizens' is still dominant, but the view that residents are 'customers' has made its way into political rhetoric. This has come mainly from conservative political sectors, but it is also part of a more general move towards the introduction of quality systems in public service provision. Associated with the modernization of public services, there is an emphasis on improved professional training for technical and administrative staff. Plans are being introduced for administrative modernization and the de-bureaucratization of municipal services, both at national and local government levels. Thus, one of the local government measures in the National Plan for 1996–1999 introduces pilot projects for reorganizing local services, improving efficiency and enhancing the efficacy and transparency of policy-making.

DECENTRALIZATION, MARKETS AND THE NEW URBAN GOVERNANCE[16]

In spite of its constitutional autonomy, an increasingly politicized local government has been under the influence of a national political project for the creation of a sound welfare state. This project has only been in the process of implementation since the Revolution of 1974, with this late start distinguishing Portugal from its EU partners. Yet Portugal has aligned itself quickly with main trends in the rest of Europe, even if its progress in welfare state provision falls below that of many European nations (Esping-Andersen, 1990; Hansen & Silva, 2000).

One element in the process of state restructuring that has occurred since 1974 is the growth of local government financial resources. An increase in the technical capacity of municipalities took place in the first years after the 1974 Revolution, which brought rapid expenditure change for two years. Ever since 1974, the upward trend in municipal expenditure has been faster than growth in the Gross National Product. Except for 1982–1984, 1986 and 1990, when block grant transfers were reduced, the annual rate of increase in municipal spending has been greater than the GNP rise. As a result, the municipal sector has seen an increased share of national wealth.

However, one element of this trend is that growth has tended to maintain the expenditure gap between municipalities. Thus, in a regression analysis undertaken on total municipal expenditure over this period, the level of expenditure in one year was largely explained by past spending commitments ($r^2 = 0.98$). Over this period inequity levels in municipal spending altered by little. At least indirectly, increases in national income were responsible for demand growth for public goods, especially those oriented toward social and cultural goods and services.[17] This demand growth has been income-elastic at the local government level, in the sense that economic growth leads to faster public expenditure increases. For instance, between 1977 and 1989, the GNP increased eleven-fold while the rate of increase for municipal expenditure was eighteen-fold. If we consider infrastructure commitments, the differences are even greater, with local government capital expenditure rising by proportionately more than its national government counterpart (Silva, 1995, p. 182). A good reason for this is that productivity in local administration has increased more slowly than in the private sector, especially when compared to manufacturing, where it is easier to replace labor with capital. Owing to the high share personnel commitments have in public sector costs (e.g. 53% in 1980), disproportionately more resources are needed to improve public service levels.[18] Compared to the national government, local services are more labor intensive, which produces a relative price effect between these two administrative levels and between government and the private sector. Trade union struggles to equalize wages at national and local government levels have also contributed to local expenditure growth in the last two decades (Silva, 1995, p. 180).

A second set of factors generating increased demand for services is demographic growth, which has been enhanced in metropolitan areas by rural exodus, from the interior to coastal areas. In the regression analysis of expenditure growth, regression residuals for the period 1976 and 1994 reveal that 65% of all urban municipalities had positive residuals, against only 28% for rural ones (Silva, 1995, p. 183). Faster population growth has been associated with these higher increases in municipal expenditure. However, while the Pearson's correlation coefficient for municipal expenditure and population growth is positive and statistically significant, it is not particularly high, which points to complexity in expenditure growth processes. One reason is the different range of services urban and rural municipalities seek to provide. When separate analyses were undertaken for each, the correlation coefficients were larger for urban municipalities, although they were still small (r = 0.45 for 1981–1991). The reason why the degree of urbanization is responsible for uneven local expenditure growth is that the urban population is less self-sufficient in its provision of social and cultural goods than the rural population. There are also different consumption patterns in larger settlements, with more collective goods needed (e.g. car parking facilities, green spaces). It follows that urban municipalities produce and deliver more services than rural ones. As the Portuguese population has become more urban,[19] local public expenditure has risen more rapidly. A notable feature in this process is that, as more households have become less self-sufficient, new infrastructure has needed to be built. For instance, in the 1994 survey of mayors, 45% reported that social housing promotion had increased strongly after 1986 (Silva, 1995, p. 363). Similarly, in a research project concluded in 1997 (Silva et al., 1997), increased planning and land use control mechanisms were reported by most urban municipalities. For the first time, Portuguese municipalities now have to have a municipal master plan for land use planning.

Uneven pressures on municipal budgets caused by population shifts have been intensified by the changing distribution of social groups and age structures, which have differentiated demand for public services. In spite of a falling birth rate,[20] the Portuguese population has increased since the 1970s. This population growth has been accompanied by notable change in population structure. The share of the population that was elderly (+65 years) increased from 9.8% in 1970 to 11.5% in 1981 and then to 13.6% in 1991. For those of less than 15 years of age the figures fell from 28.1% in 1970, to 25.5% in 1981 and then to less than 20% in 1991 (Silva, 1995, p. 180). But social composition is not just about age balances, but is related to the balance of grades on occupational scales. This is because those who are higher on occupational scales request more services than those lower down these scales (Silva, 1995, p. 189).[21] With transformation in the Portuguese economy leading to more white-collar jobs, this factor alone has made a positive contribution to government expenditure growth (Silva, 1995, p. 389). Not surprisingly in this regard, in 1994 82% of mayors held that municipal social and cultural policies had grown strongly over the 1986–1994 period. Indeed, in a survey of local councilors in municipalities and parishes (*freguesias*) in the Lisbon Metropolitan Area conducted in 1997–1998 (Silva, 1998a), one clear message that emerged was the importance of growing intervention in social policy fields (also Silva, 1999). This is not to say that economic change has made an equal impression on municipalities, for local economic structure has influenced expenditure growth (economic sectors have different needs for local public services). The correlation coefficients computed for the 275 municipalities of Portugal between expenditure increase and economic structure are statistically significant, which corroborates the link between economic sector and government expenditure. If municipalities are subdivided into different types (by size, or urban/rural location) recorded correlation coefficients are even greater (Silva, 1995, p. 191).

Adding to these 'structural' changes in the local policy environment, local government has become increasingly politicized. Although party platforms are not undifferentiated, up to this point all political parties had adopted a pro-growth attitude toward local government. As one example, in the last medium-term national plan (1996–1999) new items were proposed for the decentralization of responsibilities from the national government to local government. This orientation by the national branches of all political parties is reflected in an ideology that favors growth in local expenditure. The Communist Party, in particular, has argued on several occasions for the reinforcement of local government finance in order to allow more services to be provided to local populations.[22] Such calls for political action are not mere rhetoric for there have been real spending increases. That said, like Ashford and associates (1976), Aiken & Depré (1980), and Aiken & Martinotti (1985), empirical evidence from total 275 municipalities confirms that there are differences between local political parties in terms of their expenditure commitments. In the mayors' survey of 1994 (Silva, 1995), 34% considered that political parties are important in the determination of local public policies. In sum, party politicization has advanced over the past 25 years but with varying impressions on local policy commitments, both across parties and between categories of municipalities. Left parties are associated with higher expenditure increases. Regression residuals from expenditure levels in different electoral periods reveal that 42% of municipalities run by left-wing parties spent more than expected against only 34% for places under right-wing control (Silva, 1995, pp. 194–195). This pattern exists for total expenditure levels and for several single policy fields; most notably, housing, urban planning and social and cultural policy (Silva, 1995, 1995a). Significantly, for municipalities that remain under the control of the same party, those governed by the left tend to have higher expenditure growth rates than those run by parties of the political right. For those municipalities that see a change in political majority, those that move from right to left experience greater growth levels than those that move from left to right (Silva, 1995, p. 195).

Parallel with increased politicization, municipalities have adopted new managerial practices associated with 'new public management' ideals (Stewart & Stoker, 1995; Keen & Scase, 1998, p. 1), although most have gone down this line slowly. In practice, although a 1997–1998 councilor survey in the Lisbon Metropolitan Area revealed that most municipalities have entered service provision partnerships with non-profit organizations and other local authorities (Silva, 1998), in most municipalities a mixture of provision styles is found. This includes contracting-out, contract management, the creation of municipal public enterprises, and other forms of business-like innovations which have described by Wolman & Davis (1980), Hoffmann-Martinot & Nevers (1989), and Clark (1994, 1995), amongst others, in contexts of fiscal pressure. In the 1994 questionnaire for mayors (Silva, 1995, p. 64), 84% said their municipality had adopted innovations in the organization of municipal services. These included the introduction of new information technology practices, flexible schedules, use of voluntary workers and part-time workers, use of leasing for different types of operations, etc. The main factor responsible for the introduction of these changes was said to be fiscal stress (53% of answers), while 45% of mayors felt that private enterprise would respond better than local government to service demands in certain fields. Since the middle of the 1980s several municipalities have privatized services or transformed them into public enterprises for these reasons.[23] Likewise, in 1998, after years of successive proposals in Parliament, a new law provided a constitutional position for municipal enterprises.[24] In the 1994 survey, 82% of the mayors considered private production of certain municipal services to be advantageous and were considering 'privatizing' part of their services. About half (43%) considered it appropriate to create municipal enterprises. Contracting-out is also common nowadays, with 61% of mayors indicating that their municipalities regularly contract with private enterprises to produce municipal services (e.g. school transport, sewage infrastructure, road building and maintenance, cultural activities, industrial parks, and markets). Public-private partnerships are a generalized practice in almost all policy areas, although most notably for social policy and local economic development. Associations between public enterprises and private companies go back to the beginning of 1980s,[25] with participation in public-public partnerships also becoming common, usually with a public enterprise.[26] In some of these partnerships, municipalities retain the majority of invested capital.[27] Public participation and empowerment strategies are also an important development in municipal planning and policy formulation, as recent examples of strategic planning illustrate (Silva, 1997, pp. 68–73). The decentralization of municipal competencies to parish councils is becoming a more generalized practice.[28] Despite the general pattern of slight public involvement in municipal affairs (as seen in low electoral turnout and participation in daily management issues in local government), in recent years the decentralization of responsibilities to parishes has been affected by growing pressure from residents and citizens, as well as from opposition councilors, who express concern about citizen rights. It is possible for a group of citizens to apply directly to the court (*acção popular*) to stop municipal decisions, with several instances of this happening already.[29] This is another indication of citizen pressure on local councilors.

The ability to decentralize to parishes in this way is eased by there being no single model of municipal organization in Portugal. Rather there are combinations of different ones, associating old forms of decision-making with new business-oriented approaches. Overall, new management models[30] comprise a move from bureaucratic management forms, involving direct, hierarchical and centralized control systems, to more flexible structures, including the creation of municipal enterprises and decentralized control mechanisms. There has not been a replacement of the traditional 'bureaucratic' model, although new processes have met with some resistance from routinized bureaucratic practices and from local government managers and some local politicians.[31] No doubt these changes have influenced the growth profile of municipalities and were not simply consequences of a liberal political orientation. This is a trend informed chiefly by a critique of the wider state in general and not so much of local government in Portugal. It is somehow a soft version of the new right thinking envisaging a return of local government to its original character of organizer and facilitator, but not necessarily direct provider of local public services. Underpinning this development is also the idea that private supply has more advantages than the usually inefficient and expensive public supplier.

CONCLUSION

This chapter has examined different factors responsible for local government expenditure growth in Portugal since 1974: a growing role in the regulation process, cut in spite of being low, new and enlarged responsibilities, demographic growth, changes in social structure, in resource distribution and in urbanization. The revision of local government Acts that took place in 1998–1999 has reinforced the path of change toward more complex organizational forms. Personal income increases and a pro-growth political majority in most local authorities, as well as in the national government, can be linked to a traditionally lower level of welfare provision in Portuguese society, alongside recognition that a modern democratic state responds positively to growing citizen service demands. All parties in opposition have criticized the then national government for not transferring more resources to local government, and this has become an annual issue during budget discussions in Parliament. It seems clear that politicians have a clear perception that growing citizen demands require more and better working conditions. The Communist Party, very noticeably, calls for more resources for local government in order to improve overall quality of life for the people (PCP, 1997). All the other parties regularly support similar views about the importance of enabling local government responsiveness.

Although there has been a shift towards public-private partnerships, privatization and the contracting out of municipal services, readers should not associate this with factors like middle class taxpayer revolts, organized anti-tax groups or cutbacks in intergovernmental grants, as has been reported in other research studies (Clark et al., 1985).[32] These processes seem to be some way off the Portuguese political agenda. Retrenchment is not a potent process in Portuguese local government, despite cross-party disparity in commitments to expenditure increases, and irrespective of change in mechanisms of provision for local services. The minimalist view of an enabling council, instead of a direct service provider, still has limited expression in Portugal.

What this chapter has sought to bring about is that this is the case in spite of a growing politicization of local government that has reinforced the role of parties and politicians in local politics. In good measure, the recent introduction of democratic government in Portugal, coming as it did at a time of rapid growth in demand for public services, eased the institutionalization of coordinating mechanisms between local government and civil society. Parallel to local government politicization what developed was a process in which public policy was carried out in collaboration with private actors, which has helped configured the rise of a new form of urban governance. Municipalities are working more and more through a network of partnerships, contracts and mechanisms of influence in order to achieve desired service delivery levels. As a consequence urban governance is becomimg increasingly differentiated across municipalities, despite the introduction of organizational forms to being used to try to cover service provision shortfalls. Yet, the specificity of local government organizations, being subject to 'external' political control through local elections, plus the need to comply with national statutory obligations, is creating tension between bureaucratization and centralized control on one side and the need to introduce more flexible structures on the other.

Local government is increasingly using new instruments to enhance its capacity to act. This is being undertaken in a context of overall expenditure growth, but at the same time with retrenchment in direct public sector intervention. That said, municipalities differ in the extent and nature of change toward new management models. Indeed, changes experienced by local government in the last two decades involve continuities as well as moves forwards and backwards. As a consequence, while still seeing growth in traditional structures, policies and expenditure categories, local government is becoming more of a public-private affair in some policy areas, with market-oriented models of urban policy-making dominating. In sum, local government is being transformed by a strong decentralization movement in two directions: one from the national government, which is reinforcing the competencies and resources of municipalities, the other toward the decentralization of municipal services to markets, quasi-markets, public-private partnerships, and even parishes. The former of these trends has been apparent for the last 25 years, the latter over the last decade. The first movement is the result of the extension of democracy, whereas the second comes from the pace of demand for increased services. Together these trends are responsible for the growth of the local government system in Portugal.

NOTES

1. This research was partly developed in the Project SSPS/C/PCL/96, *Poder Local e Polìticas Sociais*, with the support of FCT-Fundação para a Ciência e Tecnologia and MSSS-Ministério da Solidariedade e Segurança Social. An earlier version was presented at the World Sociology Congress, Montreal, July-August 1998, with the support of Fundação Calouste Gulbenkian, Lisbon.

2. This means that on average Portuguese municipalities are amongst the largest in Europe, with mean average populations greater than in Spain, France, Italy, Germany, Denmark or Sweden (Stewart & Stoker, 1995, p. 253; Pratchett & Wilson, 1996, p. †13).

3. Municipalities cannot establish new taxes and, although they can raise revenue through fees, this has been controversial. In 1999, for instance, the fees collected by the municipality of Mafra were decreed by the *Tribunal de Contas* to be illegal. Only under the 1999 Local Government Act have municipalities been able to determine without restrictions the rate they set for some local taxes. In 1997, 45% of local government revenue came from the national government.

4. Municipalities have two boards – one of which is deliberative (municipal assembly), the other having an executive function (*câmara municipal*). Both boards are elected directly.

5. In 1994, the Social Democrat Party (PSD) announced that its councilors were not allowed to take a full-time post on an executive if it is a minority party. The official party line favors reform of electoral law, away from proportional representation toward a system that makes majority rule more likely (*câmara municipal*). By contrast, the Communist Party has continued its policy orientation of working with all parties 'interested in the solution of the local problems' (PCP, 1985, p. 174; also PCP, 1988).

6. In the city of Setúbal, around 92% of 1983–1985 decisions were taken unanimously on the executive board (*O Jornal* December 13, 1985). In a questionnaire distributed to mayors in 1994, Silva (1995, p. 43) found that in 45% of municipalities, the main political documents (such as the Municipal Activity Plan and the Budget) were adopted by a unanimous vote in almost all years, although this was more likely in rural than in urban municipalities.

7. Even the Communist Party, in an internal document, has noted that it is important to include in the party list people who are not affiliate with the Party but who are interested in local problems (PCP, 1988, p. 178).

8. The Communist Party regularly organizes a 'National Conference on Local Government', in which experiences are presented and policy guidelines defined (e.g. PCP 1985).

9. For example, after the PSD decided not to allow its local representatives to accept full-time local government posts if the Party is in opposition, this was ignored in several municipalities (*Público*, January 1994).

10. The questionnaire was sent to mayors and had a 26% return rate. The sample structure reflects patterns in the local government system in terms of the size, rural/urban/metropolitan location of municipalities and political party balance.

11. This situation has changed recently. Under Law 95/99, of 17 July 1999, municipalities are now able to appoint a 'General Manager' (*Director-Geral*).

12. Some parties prepare a national manifesto for local elections, in which the principles and guidelines that are to be followed by party members in municipalities or parishes are spelt out. As an illustration of the detail these documents can specify, for the 1989 election, the document prepared by the Communist Party had around 80 pages on this issue (PCP, 1988, pp. 107–190).

13. Recent examples of this include the municipalities of Amadora and Vila Franca de Xira, both in the Lisbon Metropolitan Area, which changed from Communist Party to Socialist Party control after the local election of December 1997.

14. Law 95/99 introduced the figure of a 'Municipal Administrator'.

15. The first local referendum took place in 1998 in the Viana do Castelo in northern Portugal. The second occurred in the south, in Tavira, in 1999. More referenda of this kind have recently been announced, related to issues like constructing a new car park and a seaside road. Examples of informal local consultations include one on the same day as the 1999 national election, with local residents asked whether they wanted to be integrated with the nearby municipality where most of them work (*Público*, 19 September 1999).

16. In this section, results from an analysis of municipal expenditure for the period after 1974 is presented. The lack of relevant data to measure theorized processes sensitively justifies using simple quantitative tests, such as Pearson's correlation and simple

regression equations. The size of local authorities in Portugal, and the fact the data concerns all the municipal territory, not merely its urban centre, means some results may hold more clearly for some municipalities than others (i.e. some municipalities are 'all urban' others are not). This undoubtedly lessens the statistical significance of the results obtained. Also important is the limitation of using a few measures to 'explain' a complex and temporally shifting reality.

17. According to *Wagner's Law*, increased private consumption requires improved public services (Baumol, 1967). For instance, more cars require more road infrastructure. In Portugal, the number of private cars increased between 1974 and 1991 by 260%, from approximately one million to 3.7 million vehicles (Silva, 1995, p. 239).

18. Although it is difficult to evaluate effects precisely, the type of activities carried out by local government, like land use planning, cultural services, education and social services, have less opportunities to replace labor with capital than most private sector activities.

19. In 1970, 64% of the Portuguese population lived in urban municipalities, in 1981, 68%, and in 1991, 70%.

20. 19.6% in 1974 to 11.8% in 1991

21. For instance, there has been a major increase in the school population, due to the extension of compulsory schooling from four years in 1974 to nine years by 1987/88. This was associated with increased responsibilities for local government, so yielding obvious pressure for greater local public expenditure.

22. One example of this was the Parliamentary speech made by Luís Sá (a Communist Deputy) in the debate on the Local Government Finance Act (June 3, 1998).

23. In 1991 a member of PSD Government proposed that municipalities should only have a co-ordination role, leaving the production of services to experts (*Expresso*, 1991). Four of the first examples of this movement are the municipality of Covilhã (Centre Region), which advocated in 1989 the 'privatization' of urban services, and the municipalities of Maia (North), Lagoa (South) and Mafra (Lisbon Metropolitan Area). In Lisbon, as early as 1991, even the recently elected Socialist Party mayor (in a coalition with the Communist Party) was in favor of concessions to the private sector for certain municipal functions, such as green space maintenance (*Jornal de Notícias* January, 1991). He also considered public-private partnerships to be a fundamental instrument for municipal service provision (Sampaio, 1993, p. 22).

24. This was Law 58/98, of 18 August 1998. Less than two months after the publication of this Act, several municipalities were reported to be creating this type of organization. This was the case for Braga, VN Famalicão and Guimarães. In the municipality of Braga, three municipal enterprises had been created by December 1998 (*GERE-Empresa de Águas, Efluentes e Resíduos Sólidos de Braga, TUB-Empresa de Transportes Urbanos de Braga* and *PEB-Empresa do Parque de Exposições de Braga*).

25. Some of the earliest examples of the new 'generation' of public-private partnerships are: the *Sociedade Ribeira-Pêra*, created in 1984, in Castanheira de Pêra, in which the municipality owned 39% of the capital (Henriques et al. 1986, p. 123); the enterprise ENASEL SA, which joined with the municipality and a public enterprise to promote tourism in an area associated with hunting; and, enterprises created for the production of hydro-electricity in the north of the country (*Diário de Notícias* 1989; *Independente* 1990; *Público* 1991; *Jornal de Notícias* 1991). Ambelis, the Lisbon Development Agency, created by the city of Lisbon but increasingly with a metropolitan focus, is another example of a partnerships in the economic sphere (Syrett & Silva, 1999). For partnership in social

services, see Silva (1998a). Examples of public-private partnerships in housing can be seen in Silva & Hoggart (1998). One of the most recent examples is the creation of the *Sociedade de Desenvolvimento da Frente Ribeirinha Norte e Atlântica-Turiscost SA*, in which the State has 69% and the municipality of Almada 31% of the capital.

26. A paradigmatic case is the public enterprise to be created in the near future that will join the municipalities of Lisbon and Loures with the state-owned company Parque Expo (the latter was the company that promoted the Lisbon World Expo 98). The company formed by these three partners will be in charge of urban management of the Expo area.

27. A recent example is the 1999 legislation (Law 176/99) that allows the state to alienate, in favor of municipalities, its majority capital position in joint venture companies created after 1993 (Law 379/93) to run inter-municipal environmental systems (water, waste, etc.). Later this law was cancelled, but the issue is on the table and will soon be discussed again.

28. For example, the municipality of Amadora has recently made a general agreement with all its parishes (*Público* March 8, 1999). As this involves the transfer of resources, this complicates the puzzle of knowing how many services the municipality actually delivers. There are, for example, municipalities where relations between municipalities and parishes are not formally established in a protocol, with arrangements more commonly the object of one-off decisions, which are open to discriminatory practices (*Público* January 13, 1999). For an analysis of decentralization from municipalities to parishes in the field of social policy see Silva (1998a). The impetus to decentralize services to parishes comes from both parish and municipal levels. In part this reflects a vision that services are more appropriately provided at a more local level (on efficiency and efficacy grounds). In part it represents a desire by some parish councils to increase their official sphere of competence over service provision.

29. As part of the 'Citizens Guarantees and Liberties Principle' of the 1976 Constitution, it is a citizen's right to appeal against such decisions. What is a recent change, is an increase in citizen usage of this constitutional provision.

30. As Stoker (1999, p. 2) and Lowndes (1999, p. 22) note, 'new management' has proved a difficult concept to define. The phrase appears to contain different and contradictory ideas and practices.

31. In recognition of this process the 1999 Local Government Act (Law 169/99) gave more authority to the Municipal Assembly over scrutinizing municipal involvement in new organizational forms, including associations and federations, public and private enterprises, co-operatives, foundations or other organizations in which the municipality is a stakeholder.

32. In spite of the fact that EU Structural Funds continue to fuel the system with substantial resources, there are signs from the national government that municipalities will have to find alternative and complementary funding. One example of this sentiment is the statement by the Ministry of Environment that municipalities will have to find additional funds from civil society for their environmental rehabilitation policies (*Público* May 4, 1999). Another factor that favors the adoption of fiscal restraint is the need to meet European Monetary Union debt criteria. While the national government and the social security system were able to reduce debt by 8% between 1995 and 1998, over the same period local government debt continued to increase (*Público* February 23, 1999).

REFERENCES

Aiken, M., & Martinotti, G. (1985). The Impact of Political and Urban Systems on Municipal Expenditure. In: T. N. Clark (Ed.), Urban Innovations as Response to Urban Fiscal Strain, (33–57). Berlin: Verlag Europaische Perspektiven.
Aiken, M., & Depré, R. (1980). Policy and Politics in Belgian Cities. *Policy and Politics 8*, 73–106.
Amaral, D. F. (1993). *Curso de Direito Administrativo*. Coimbra: Livraria Almedina.
Ashford, D., Berne, R., & Schramm, R. (1976). The Expenditure-Financing Decision in British Local Government. *Policy and Politics 5*, 5–24.
Baumol, W. J. (1967). Macroeconomics of Unbalanced Growth: the Anatomy of Urban Crisis. *American Economic Review 57*, 415–426.
Caetano, M. (1983). *Manual de Direito Administrativo*. Coimbra: Livraria Almedina.
Clark, T. N., Hellstern, G. M., & Martinotti, G. (Eds.). (1985). *Urban Innovations as Response to Urban Fiscal Strain*. Berlin: Verlag Europaische Perspektiven.
Clark, T. N. (Ed.). (1994). *Urban Innovation*. London: Sage.
Clark, T. N. (Ed.). (1995). Les Stratégies de lInnovation dans les Collectivités Territoriales: Leçons Internationales. *Politiques et Management Public 13*(3), 29–61.
Cochrane, A. (1993). *Whatever Happened to Local Government?* Buckingham: Open University Press.
Duncan, S., & Goodwin, M. (1988). *The Local State and Uneven Development: Behind the Local Government Crisis*. Cambridge: Polity.
Esping-Andersen, G. (1990). *The Three Worlds of Welfare Capitalism*. Cambridge: Polity.
Hansen, F., & Silva, C. N. (2000). Transformation in the Welfare States After World War II: The Case of Portugal and Denmark. *Environment and Planning C: Government and Policy* (forthcoming)
Henriques, J. M., & Neves, O. (1986). Castanheira de Pêra: Uma Via para o Desenvolvimento Regional Endógeno? *Sociedade e Território 2*(4), 116–125.
Hoffmann-Martinot, V., & Nevers, J-Y. (1989). French Local Policy Change in a Period of Austerity. In: S. Clarke (Ed.), *Urban Innovation and Autonomy*, (182–213). Thousand Oaks, California: Sage.
Judge, D., Stoker, G., & Wolman, H. (Eds.). (1995). *Theories of Urban Politics*. London: Sage.
Keen, L., & Scase, R. (1998). *Local Government Management*. Buckingham: Open University Press.
Lowndes, V. (1999). Management Change in Local Governance. In: G. Stoker (Ed.), *The New Management of British Local Governance*, (22–39). London: Macmillan, 22–39.
Miranda, R., & Walzer, N. (1994). Growth and Decline of City Government. In: T.N. Clark (Ed.), *Urban Innovation*, (146–166). London: Sage.
National press: *Diário de Notícias, Jornal de Notícias, Público, O Jornal, Expresso, Independente* (across several years).
Newton, K. (1974). Community Performance in Britain. *Current Sociology 22*, 49–86.
Oliveira, A. C. (1993). *Direito das Autarquias Locais*. Coimbra: Coimbra Editora.
PCP (1985). *Reforçar o Poder Local Democrático, Melhorar a Vida das Populações*. Lisbon: Edições Avante.
PCP (1988). *Encontro Nacional do PCP sobre o Poder Local: Poder Local Presente e Futuro*. Lisbon: Edições Avante.
PCP (1997). *Conferência Nacional do PCP. O Poder Local e as Eleições Autárquicas — Intervenção de Carlos Carvalhas*.

Pierre, J. (1998). Public-Private Partnerships and Urban Governance. In: J. Pierre (Ed.), *Partnerships in Urban Governance: European and American Experience*, (1–10). Basingstoke: Macmillan.
Pratchett, L., & Wilson, D., (Eds.). (1996). *Local Democracy and Local Government*. London: MacMillan.
Rallings, C., Temple, M., & Thrasher, M. (1996). Participation in Local Elections. In: L. Pratchett & D. Wilson (Ed.), *Local Democracy and Local Government*, (62–84). London. MacMillan.
Sá, L. (1991). *Razões do Poder Local*. Lisbon: Editorial Caminho.
Sampaio, J. (1993). Discurso do Presidente da Câmara Municipal de Lisboa. In: C. M. Oeiras (Ed.), *Proceedings of Seminário Internacional sobre o Poder Local*, 18–24.
Silva, C. N. (1995). *Poder Local e Território: Análise Geográfica das Políticas Municipais, 1974–94*. Lisbon, Universidade de Lisboa.
Silva, C. N. (1995a). Autarquias Locais e Gestão do Território: Que Diferença Faz o Partido Político? *Finisterra 59/60*, 99–120.
Silva, C. N. (1997). Planeamento e Participação do Público. In: APEA Associação Portuguesa de Engenheiros do Ambiente (Ed.), *Workshop Sobre Participação Pública*, (68–73), Lisbon.
Silva, C. N. (1998). Local Finance in Portugal: Recent Proposals and Consequences for Urban Management. *Environment and Planning C: Government and Policy 16*, 411–421.
Silva, C. N. (1998a). *Poder Local e Políticas Sociais*. Lisbon: unpublished research report, Centro de Estudos Geográficos, Universidade de Lisboa.
Silva, C. N. (1999). Local Government, Ethnicity and Social Exclusion in Portugal. In: Khakee, P. Somma & H. Thomas (Eds.), *Urban Renewal, Ethnicity and Social Exclusion in Europe*, (126–147). Aldershot: Ashgate.
Silva, C. N., & Hoggart, K. (1998). Parcerias Público-Privado nas Políticas de Habitação em Portugal e no Reino Unido. *Inforgeo — Revista da Associação Portuguesa de Geógrafos* 12/13, 363–371.
Silva, C. N. et al. (1997). *Metodologia e Indicadores de Avaliação de PDM*. Lisbon: unpublished research report, Centro de Estudos Geográficos, Universidade de Lisboa.
STAPE various years. *Resultados eleitorais*. Lisbon: Secretariado Técnico para os Assuntos do Processo Eleitoral.
Stewart, J., & Stoker, G. (Eds.). (1995). *Local Government in the 90s*. London: Macmillan.
Stoker, G. (Ed.). (1999). *The New Management of British Local Governance*. London: Macmillan.
Syrett, S., & Silva, C. N. (1999). New Institutions for New Forms of Governance: Regional Development Agencies in Portugal. Paper presented in Regional Studies Association International Conference, 18–21 September 1999, Bilbao.
Tarschys, D. (1975). The Growth of Public Expenditures: Nine Modes of Explanation. *Scandinavian Political Studies 10*, 9–31.
Walsh, K. (1988). Fiscal crisis and stress: origins and implications. In: R. Paddison & S. Bailey (Eds.), *Local Government Finance*, (30–51). London: Routledge.
Walsh, K. (1995). *Public Services and Market Mechanisms: Competition, Contracting and the New Public Management*. London: Macmillan.
Wolman, H., & Davis, B. (1980). *Local Government Strategies to Cope with Fiscal Pressure*. Washington DC: The Urban Institute.

10. ENHANCING LOCAL FISCAL AUTONOMY: THE JAPANESE CASE WITH COMPARATIVE REFERENCE TO SOUTH KOREA AND THE UNITED STATES

Yoshiaki Kobayashi

When the total deficits compiled by central and local governments are combined, Japan's total national debt reaches 101.8% of the Japanese Gross Domestic Product. In other words, Japan's balance of payments exceeds the nation's GDP. As a result, a fair portion of all tax revenues simply goes toward paying interest on debt. Management of Japan's roughly 3,400 local administrative units is typically inferior to what is acceptable in private enterprise. This poor management is one of the primary causes underlying Japan's fiscal problems.

Although each municipality provides fairly standard services, the national government, through what are called 'ordinary allocation tax funds', covers the difference between localities' standard revenues and the cost of the numerous services that are paid for by the municipality. However, if, for example, local corporate taxes were increased, then the local allocation tax would be reduced by the same amount. Therefore, most municipalities not only fail to cut their expenditure, but even indulge in constructing such facilities as gyms and music halls, which are typically an unaffordable extravagance for the local jurisdictions (Sakata, 1998). This is a trend we should expect to continue into the foreseeable future.

However, even at the national government level, revenues fall short, and problems of subsidizing municipalities will continue into the future (Hayashi, 1999). Today, we have a golden opportunity to reconsider local government management practices to rid ourselves of the inefficiencies and waste inherent in them. To better tackle this problem, we ought to begin by re-examining the relationship between the central and local governments in Japan.

To consider this problem, this chapter is broken down into three parts. In the first part, the results of a survey of local government leaders responsible for fiscal policy in Japan, South Korea, and the USA are examined (Appendix D). Comparing the answers provided by these leaders helps us better understand the problems of Japan. The highly decentralized United States (Appendix C) is furthest from the centralized Japanese case (Appendix A). By comparing the level of fiscal responsibility witnessed in these two countries, it will be shown that a high level of centralization has helped bring about Japan's local governmental fiscal crises. South Korea (Appendix B) offers a different comparison. Like Japan, South Korea has a highly centralized system. As such, a comparison of Korea with the USA offers the ability to assess how generalizable conclusions are on the impact of centralization on local government fiscal management. However, differences in fiscal responsibility exist between Japan and Korea. Korean municipalities are forced to demonstrate greater accountability than their Japanese counterparts. This is related to their greater ability to raise revenue and their tighter links to the center (Korean municipalities directly collect more categories of tax than Japanese municipalities, and depend on national government subsidies for only about one-quarter of their revenue). These differences highlight that the root of many fiscal management problems in Japanese local government lie in Japan's overly centralized governmental fiscal system, alongside a lack of accountability tying the center and municipalities.

One of the lessons we can learn from the Japan-Korea comparison is that greater consolidation of local governments may help reduce problems. In the second part of the chapter, a simulation exercise is presented to demonstrate what fiscal effects consolidation might have in Japan. Finally, in the last part of the chapter, a number of potential reforms that could begin to rein in the terrible problems Japan's municipalities face are presented. In particular it is recommended that Japan seriously alters its system of local revenue subsidies, introduces a minimum local taxation rate and greater local fiscal responsibility, promotes citizen participation in policy-making and creates standardized municipalities through consolidation.

LOCAL GOVERNMENTAL FISCAL POLICY: RELIANCE ON THE NATIONAL GOVERNMENT

Japanese local governments, like those in South Korea, are extremely dependent on the national government for their revenue. Because they are thereby less accountable locally for the funds they spend, local governments tend toward irresponsible fiscal management, especially when compared with municipalities in the USA. This lack of accountability is particularly noteworthy when we compare Japan with South Korea. In the latter case, municipalities are given greater autonomy over their ability to acquire local revenues and display greater long-term responsibility in their fiscal management.

In Japan the largest problem municipalities face in financing programs is that they depend so much on funding from national revenues. Japanese cities receive 23% of their funds from the center, and for smaller towns and villages the figure rises to 48%. Yet national regulation of municipal utilization of independent revenue sources, alongside the reallocation of monies from the center to municipalities, through policies like the local allocation tax, do make it possible to standardize local services, even in poorer areas. However, to the extent that this system is administered by the national government by distributing funds to the municipalities, it forces municipalities to lobby constantly and reduces incentives local government might have to increase revenues on their own.

In contrast to the strong national-local government revenue link that we see in Japan, in the United States there tends to be more direct lobbying by local residents to their city governments. Because of the link between budget increases and tax increases, where possible taxpayers usually want to find ways to cut the budget. Yet beneficiaries of governmental spending have a great stake in maintaining and increasing social welfare expenditures. The degree to which taxpayers and government spending beneficiaries hold power vis-à-vis one another is thereby a critical factor shaping the administration and financing of local spending on governmental services (see Takayose, 1998). Whereas, in the relatively centralized Japanese system, municipal revenue problems stem from issues surrounding the national-local government relationship, in the comparatively decentralized USA, control of revenues depends heavily on the nature of power within municipalities (see Kobayashi, 1998).

The survey data presented in this chapter enables us to see the extent to which these generally held beliefs are true. It does so, first, by looking at the relationship between local fiscal planners and higher-tier governments. To what extent are problems seen to be due to insufficient national government or prefectural/state government funding, to revenue decline, or to an inability to secure loans to cover project costs? If funds from higher level governments (such as prefectures and large cities) play a large part in local government finance, there should be significant awareness of the magnitude of this problem among local leaders. In this regard, the survey reveals noticeable differences in the responses of local leaders in the more centralized Japanese and South Korean cases than in the USA. For example, while roughly 60% of local leaders in Japan and South Korea noted that insufficient national government finance was 'extremely important' in local government financial circumstances, just short of 5% of U.S. respondents found insufficient federal financing was 'extremely important'. If we include those who thought national-level funding was 'quite important', nearly every local fiscal leader held that insufficient national government support was an important factor in Japan and South Korea, whereas in the USA the figure still only reached about 20%. Although the strong links between U.S. local and state governments might account for some of this disparity, only about half of the U.S. respondents held that insufficient financing by state governments is responsible for local fiscal problems. Similarly, not much over 10% of U.S. respondents cited bond initiative failures as important factors in local fiscal problems, with the figure for Japan some four times greater, while more than half of South Korean informants saw this as an important factor.

Unlike in the USA, local leaders in Japan and South Korea perceive higher-tier government subsidies as a given. When there are insufficient funds due to bonds being turned down or rejected by higher level governments (such as prefectures), the impact on municipalities is quite severe. If subsidies are not increased at this time, then, depending on economic conditions, there may be great concern that there will be insufficient financing for local functions. In this circumstance, municipalities appeal to the national government. Noticeably, local leaders have greater interest in extracting funds from the national government at this time than in maintaining discretion over local management.[1] This sentiment is heightened by local leaders' images that citizens react to budget cuts more aversely in Japan and South Korea than do leaders themselves. Thus, whereas about 55% of local leaders in Japan believe that citizens would prefer to see higher levels municipal spending, the figure for leaders themselves is under 40%. Interestingly, this relationship is reversed in the USA and South Korea. In the former, around 30% of local leaders would like to expand municipal spending but only 10% believe this would be supported by the citizenry, with equivalent proportions for South Korea of just over and just under 50%. Irrespective of the precise balance between leaders' images of citizen preferences and their own preferences, an obvious disparity exists between the USA and the other two nations. This is that around 50% of local leaders in Japan and South Korea operate in an environment in which they believe citizens want higher local expenditure, compared with just 10% of U.S. municipal leaders who see themselves in this position.

At the same time, the dependence of municipalities on the national government is unlikely to change much, even if fiscal conditions become significantly more grave. This is seen if we look at the policies municipalities choose for the sake of their fiscal health. Here we find great differences between Japan, the USA and South Korea when we compare preferences over four possible fiscal reconstruction policies – introducing new revenue sources, increasing tax revenues, transferring funds, and increasing long-term loans. In Japan, especially when compared to the USA and South Korea, municipal leaders are prone to want extra funds from higher-tier authorities (four-fifths) or long-term loans to stabilize revenues (almost 70%). Those who wish to introduce new sources of revenue (30%) or increase taxes (less than 30%) are in the minority. The trend is to favor measures that do not increase the burden on citizens living in the area at present. By contrast, in the United States, instead of preferring the long term bond option (less than 30%), the trend is toward favoring policies that place responsibility on the current residents (more than three-fifths would wish to introduce new revenue sources, with more than half wishing to raise taxes). In large measure, because they tend to cover a larger subsection of the country, South Korean municipalities, when compared to those in Japan, are given greater autonomy and responsibility in raising funds on their own. As a result Korean local governments tend to avoid short-term solutions that may hamper their finances for many years to come (less than 30% of local leaders would want to increase long term loans, with about half preferring both broadening the revenue base and increasing taxes). It would appear that, because Japanese municipalities have little accountability in the management of their funds, they tend to favor potentially irresponsible fiscal management decisions, without serious regard to their long-term (negative) implications.

Though it may be difficult to introduce new sources of revenue in Japan, it cannot be denied that municipalities are lagging in their tax-collecting efforts. In Japan, given the particular local revenue system in which municipalities operate, local officials work primarily toward reforming deficiencies in revenue and avoiding increases in current residents' burden. For this reason, Japanese municipalities are prone to increase loans and dismantle programs – perhaps in an effort to hold off a sudden rise in fiscal strain – but they do not seek to reform the basis of their shaky financial system. As a result, Japan's local governments keep piling on debt. Since the Oil Shocks of the 1970s, Japan has dealt with the burden of its national and local debt by issuing bonds. The time has come to introduce a policy to increase revenue in order to pay off the tremendous burden future generations will otherwise have to account for.

FISCAL MONITORING AND PUBLIC INFORMATION

To summarize the previous section, because the bulk of its municipalities depend so heavily on national subsidies, and do not develop their own measures for increasing revenue, Japan has built up long-term fiscal policies that are not ideal. The fact that the center is the primary provider of funding for municipalities encourages fiscal carelessness. This is made even more apparent as we consider municipal fiscal monitoring. Municipalities in the decentralized U.S. system tend toward much greater and more sophisticated vigilance in monitoring their finance than the more centralized systems in Japan and South Korea. The importance of creating a sense of local accountability is made even more apparent when we see that in Korean municipalities, where local leaders have traditionally been more accountable for government finances, fiscal monitoring occurs at a greater rate than in Japan, where such accountability is far more limited. That said, the emphasis on generalist training of officials in Japan ought to make us wary of quickly introducing a more decentralized apparatus in the country.

Compared to the USA, Japan's financial reporting and inspection systems are not well equipped. In the USA, about 15% of municipalities engage in computer-based financial checks at least once a week, while under 5% engage in as little as four or fewer investigations in a year. In South Korea, roughly half of all local governments monitor financial affairs once a month. The Japanese system is less vigilant in financial monitoring. Only slightly more than half of all local governments undertake financial checks more than four times a year. Once again, comparison highlights the impact of great centralization on local fiscal responsibility. In the decentralized U.S. system, municipalities are the most hands-on in their fiscal monitoring, with Japanese municipalities being the least pro-active. South Korea, like Japan, has a fairly centralized system, but, because its local leaders were nominated to their posts by the center until the mid-1990s, Korean municipal leaders tend to believe they have a greater stake in carefully maintaining their finances; they monitor fiscal affairs more thoroughly than municipalities in Japan. To explain this point more clearly, South Korean mayors were effectively national bureaucrats, rather than elected officials. The fact that future postings and promotions were to be based on performance in a current post inspired more responsibility than even the electoral mechanism does now. In truth, the lack of monitoring that occurs in Japan is often not due to laziness or irresponsibility, but is an assertion of autonomy on the part local officials.

To emphasize this point, when we look at accounting and financial reports, many municipalities in Japan and South Korea manage their accounts within the guidelines specified to them. In contrast, for U.S. municipalities, only a relatively small number merely follow the bare minimum stipulations of the law (just over 10% of local leaders in the USA indicate that they merely comply with the law in the U.S., compared with almost 50% in Japan and South Korea). Compared to the USA, Japanese municipalities are not sufficiently mindful of monitoring their financial affairs. As an indication of just how little weight is attached to accountability through fiscal monitoring in Japanese municipalities, it is noteworthy that while Japan is a technologically advanced society, municipalities have not reached the point of actively using computers to monitor their affairs.

When we look at this condition, we can easily conjecture that additional problems would arise if full decentralization were introduced in Japan (or South Korea). While it would perhaps overstep the bounds of this analysis to say that the U.S. local government system is the best, it would not be an overstatement to say that Japan and South Korea could learn a great deal about the merits of decentralization from the USA. In particular, Japan might learn ways to escape its current fiscal problems that result from the nature of national-local government relations.

In order to deal with these problems, specialist training is needed for those who work in local government fiscal management. Despite the fact that economic fluctuations have an enormous impact on fiscal management, few municipalities systematically research economic trends in Japan and South Korea: only about 5% of municipalities do so in Japan and about 15% do in Korea, while the figure in the U.S. is around 35%. It is not surprising that problems like this arise in societies like Japan, where emphasis is usually placed on largely generalist training for government (and non-government) personnel. Given the disparities in research on economic trends, it is reasonable to conclude that Japanese and South Korean municipalities can not adapt well to the dynamic economic fluctuations that they frequently face.

In short, despite the tremendous problems that grow out of placing too much dependence on the national government, there is an insufficient infrastructural base for local government to function autonomously in Japan and South Korea. Within today's system of lax local government administration, separation of local government from national government control could lead to a decline in the quality of administrative services and, ultimately, harm for local residents.

Further illustration of this point comes if we consider the problem of public information. Irrespective of internal financial monitoring systems, if there is a good system for collecting public information, this could lead to more awareness of resident appeals over local government finances. Words like 'decentralization' in Japan – and 'informationalization' across the developed world – have recently come to the fore to describe important changes that have occurred in local government. Integral to these changes, serious advances have been made in information technologies, as seen in the spread of the Internet. However, when we look at Japanese and South Korean local government, the use of information technology is relatively poor. Allied with this, and most evidently when comparison is made with the USA, there seem to be fewer opportunities for using IT for reporting in the local fiscal realm. In both Japan and South Korea, many commentators are pressing governments to take advantage of information technologies. In some cases there has been a response to such calls. In Japan's Mie Prefecture, for example, the goal has been to develop a systematic network for carrying out freedom-of-information requests. But a system that would fulfill such information demands is not adequately prepared because there is no well-established way to discern public opinion. Only about 20% of Japanese and South Korean municipalities are equipped with a system of public reporting, in order to acquire an understanding of public opinion. In the USA the figure is 60%. It is important that Japanese municipalities create an information system with links to public opinion. At present few municipalities have adopted such a system, although a system based on free exchange of information would probably play an important part in acquiring public views on fiscal management. The lack of free information regarding local governments' financial affairs has helped create a situation where officials engage in scandalous practices, such as reimbursement for phony business trips and the receipt of generous food expense accounts.

It goes without saying that because administrations are financed by voter taxes, voters should be able to view their municipalities' financial information. But reports on expenditure, for example, generally contain only the broadest of categories, and allow little detailed scrutiny. What little information is officially made available is often difficult to acquire in practice, as it is cloaked by restrictive viewing rules and exorbitant photocopying fees. Where voters do not have access to information over how their tax money is spent, it becomes difficult to say that the system has an advanced informational system. Indeed, given the unhealthy direction local governmental fiscal policies are moving, it is indispensable that voters be allowed the opportunity to gain such information. In short, in order to offer voters a chance to monitor local governmental behavior, a fully equipped public information system ought to be introduced.

LEGISLATURES AND POLITICAL PARTY PARTICIPATION

Heavy centralization has not only led to a weak sense of accountability in fiscal management and monitoring in Japanese and Korean municipalities, but has helped induce a similar irresponsibility in Japanese and Korean local assemblies. Assemblies in South Korea and Japan tend to engage in little monitoring of budgetary expenditure. Yet they are all too happy to become involved in the process of spending money. This lack of legislative responsibility is exacerbated by a lack of political party competition in local legislatures in both countries.

The fiscal authorities surveyed in Japan and South Korea indicated some reluctance to offer fiscal reports to their local legislatures more than once a year (more than 40% in both countries only report this regularly, with virtually none in the USA having so irregular a reporting period). Is a single fiscal report per year sufficient for local legislators who hold budget-making authority? In both Japan and South Korea, only the executive enjoys the right to introduce budget-related bills, so local legislatures' power is inherently weak as a result. Even so, in a dual representation system, legislatures are expected to check on the executive. If legislatures are happy with infrequent reports on executive actions, this rather suggests that they are not functioning as they ought to.

Interestingly, though, while Japanese local legislatures are fairly passive in their fiscal monitoring and inspections, they are extremely active in their use of the budget. Looking at what groups have influenced project decisions over the past five years in Japan, local leaders report that city legislatures have played a particularly important role (followed by higher level governments, such as the prefectural and national governments; see Table 10.1). We see a similar trend in South Korea, where city legislatures hold the greatest weight, followed by state governments, the national government, and city mayors. In short, it appears that in Japan and South Korea local legislatures play a very active role in the process of budgetary spending allocations. They like to spend without taking responsibility for monitoring what happens once these spending decisions have been made. By contrast, in the USA, much greater influence in the project allocation process is reported for city mayors and city managers than for local legislatures, with city section chiefs wielding equivalent influence to that of local legislators.

Local legislatures in Japan and South Korea are not passive monitors of local expenditure, but are active players in decision-making processes. Yet, over monitoring issues, laxity is pronounced. In Japan, slack processes of monitoring local leaders' food and travel expenses, and scandals surrounding abuses of these privileges, have recently come to light. Local legislatures have shown a genuine lack of will to find a way to fix unhealthy aspects of their financial affairs. With the decentralization of authority now taking place, the importance of local legislatures will increase (and this process is likely to speed up in Japan

Table 10.1. Political Agents that Influence Project Decisions in Municipalities in Japan, South Korea and the USA

	Japan		South Korea		USA	
	Number	Rank	Number	Rank	Number	Rank
City legislators	254	1	108	1	61	4
States/prefectures/metropolis	174	2	100	2	19	10
National government	159	3	79	3	7	13
Neighborhood associations	139	4	13	11	29	8
Mayors	134	5	79	3	89	2
Citizen groups	111	6	27	6	7	13
Business groups	109	7	3	14	53	6
City-level section chiefs	84	8	14	10	62	3
City-level finance bureau	83	9	34	5	40	7
Individual citizens	59	10	23	8	61	4
Senior citizens groups	34	11	5	12	13	11
Socially disadvantaged	33	12	1	16	*	*
Political parties [all mentions]	33	13	5	12	13	11
Civil servants	11	15*	26	7	20	9
Local media	11	15*	22	9	6	16*
Religious/church groups	0	17	2	15	2	18*
City managers	*	*	*	*	95	1

Note: For Japan, rank 14 was occupied by labor unions, which had 18 nominations.
'Socially deprived groups' was not a category in the survey of U.S. mayors. Minority groups were included and received 4 nominations, with low-income groups also receiving 2 nominations in this country. As well as the single nomination for socially deprived groups in South Korea, there were 6 (additional) mentions for low-income groups. If added to the value for socially deprived groups, this would rank this merged group as 12th, with all other categories falling by one rank if they have fewer than 7 nominations. With 2 nominations for low-income groups in the USA, this category would be ranked 18th equal.
In the USA, environmental groups were ranked 16th equal (with 7 nominations) and taxpayers' groups 18th equal (with 6), but these were not items in the surveys in the other two countries. Gay rights groups and women's groups were included in the US survey but both received no nominations.
'City manager' was not a category in the Japanese and South Korean surveys.

under the Local Autonomy Law that took effect in April 2000). But it is not clear that local legislatures are able to manage their financial affairs satisfactorily. A credible system of fiscal monitoring is clearly needed, involving referenda, ombudsmen, or external monitoring. At present, only a few cities in Japan have ombudsmen, and virtually none undertake systematic monitoring of their own fiscal affairs. What monitoring is done by the national government is restricted to projects involving national funds.

The relationship between parties and legislatures on the one hand and the executive on the other also plays an important role in creating the problems of fiscal mismanagement. In the USA, there is not always a strong connection between political parties and elected executives. In Japan and South Korea the relationship between executive candidates and political parties is clear. Thus, whereas almost 40% of local electoral candidates in the USA reported that they did not have a party affiliation, in Japan and South Korea the comparable proportion was just over 10%. At the same time, there is a marked disparity in reliance on political parties for electoral support, with more than 60% of U.S. candidates reporting that political parties did not campaign on their behalf in local elections, compared with less than 10% stating the same in Japan and South Korea. In truth, many local legislatures in South Korea are really one-party systems, with fiscal monitoring at its lowest level in one-party legislatures, although monitoring does tend to improve where two-party competition exists. Comparable problems arise in Japan, where local leaders consider numerous requests from local legislatures in shaping the budget in order to get re-elected, with this tendency growing the longer representatives are in office (Figure 10.1). This plays a large part in the problem of municipal overspending, for such executive-legislative relations undermine any system of checks and balances. Moreover, the recent trend in Japan, whereby one candidate runs for office with the support of all parties, has altered the ability of local legislatures to act as a check on elected officials. This is not unrelated to recent debates over the introduction of citizen referenda. In more than half of Japan's prefectures, for example, the governor is supported by all parties except the Communists, and elections are essentially votes of confidence rather than issue-based contests. In cities, though party endorsements are not always as explicit, a similar trend holds for mayors. Legislatures are transformed into caucuses of the governing coalition, sometimes with no members opposing the executive. Citizen attempts to use the legislature to express dissent are often frustrated. What competition exists is often between rival conservative factions, and is waged on personality rather than party or policy grounds.

POPULAR UNDERSTANDING OF THE PROBLEMS OF LOCAL FISCAL AFFAIRS

Recently in Japan and South Korea, there has been a great deal of discussion about decentralization, but little consensus has formed. In both countries quite a few municipalities feel constrained by the national government. In Ichon City in Korea, for example, social welfare services for all but the elderly and the homeless have been almost completely cut. Revenue constraints have become an important problem – all the more so because of the curtailment of subsidies

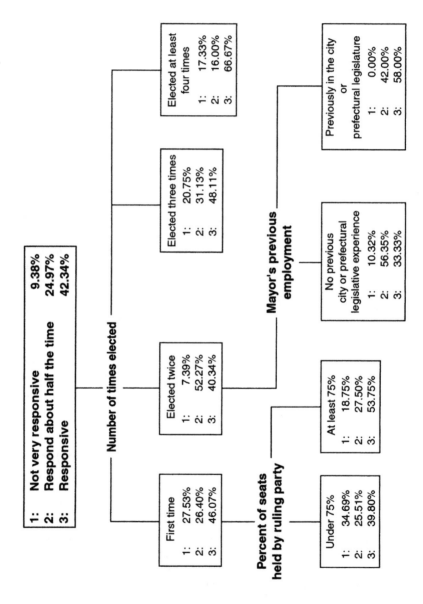

Figure 10.1: Mayoral Responses to Legislature Requests in Japan.

and the denial of new loans. In particular, the Japanese system of constant subsidization has given birth to a lax local governmental system. Moreover, even in plans to reconstruct the financial system, priority is typically given to policies favoring long-term loans that put little burden on current residents.

Given these problems, if the Japanese local government system is to move toward a more fiscally responsible operation, reform ought to focus on the system of municipal subsidies, and perhaps raising the consciousness of voters. As part of administrative reform efforts, there has been progress in reallocating and standardizing the local subsidy process, but, because there have been few suggestions for alternative forms of revenue, battles over acquiring grants have become more intense. But the process of allocating subsidies still tends toward *ad hoc* decision-making and remains deficient in collecting taxes to cover municipal spending. At present, local taxes account for only 36.6% of local revenues in Japan. Though the subsidies that take care of much of this shortfall are as often as possible distributed according to objective criteria, the distribution of funds for certain projects, including large and indivisible ones, necessarily involves subjective and political criteria (Jinno & Kaneko, 1998). At present, then, there is still need for a system that focuses as much on raising new revenues as on the equal allocation of subsidies. Compared to the USA, there is an insufficient sense of local governmental and individual responsibility over Japan's fiscal affairs. Without coming to terms with these issues it will be difficult for Japan to fix its fiscal problems.

In recent years, the debate over administrative fiscal reform and national-local government relations in Japan has grown more intense. One clear conclusion has come from these debates. This is that neither complete local freedom nor complete central control will yield a system that offers both stability and local autonomy. In addition to allowing local autonomy, it is important to maintain a system in which national and higher governments can offer some fiscal services. In this way the system can ensure a 'civil minimum' for citizens. Ultimately, then, it is vital to develop a system that enforces the principle of individual responsibility, while also providing a level of services at least equal to that needed throughout society. However, changing the relationship between Japan's national and local governments is critical to developing such a system. Rather than closing with a reiteration of the argument that comparing Japan to the USA and South Korea reveals serious failings in the heavily centralized Japanese system, which places too little emphasis on local accountability and responsibility, I would like to close with suggestions on how to alter the system that has taken such firm root in the Japanese political world.

A SIMULATION OF MUNICIPAL CONSOLIDATION

The fact that South Korean municipalities occupy a larger subsection of the country than is the case in Japan, as well as being given greater fiscal autonomy, makes for a significant difference in inducing a greater sense of accountability and responsibility in Korean fiscal management than in Japan. As such, it seems logical to suggest that consolidation of municipalities offers one essential step toward strengthening the fiscal bases of local governments. Japan has undergone two major consolidation programs in its history, both with this goal in mind. But what fiscal effects would consolidation yield? In the following sections, I consider this question through simulation models.

Standards for Consolidating Municipalities

A question to start with is, what size of municipalities would be most appropriate? This is a difficult question that is ultimately intertwined with the ability of governments to govern and serve the interests of the people being governed. In the event that there is a serious attempt at local consolidation, this must be done with an eye toward achieving the goals of administration. In the large-scale consolidation of municipalities in the late 1800s and in the early postwar years, local unification progressed with a view to serving specifically administrative aims. For example, the postwar local consolidations were premised on the assumption that 8,000 people per municipality was an appropriate size for the rational administration of the new middle school system. The postwar Shoup Report laid out conceptual guidelines for the scope of mergers, and the Kobe Conference on local administration recommended maintaining municipalities at the 8,000 level, as an appropriate size for effective governmental management.

As Kurokawa (1997) notes, while recent discussion over the scale of local consolidation has been based on likely effects on governmental efficiency, other considerations, such as the need to maintain independent regional economies, remain important as well. Kurokawa recommends respecting socioeconomic conditions and creating municipalities with populations of approximately 300,000, to make possible the maintenance of relatively autonomous local economies.

With these considerations in mind, I have drawn up three models, based on population and fiscal capability, by which Japan might choose to consolidate its municipalities. The population levels are based on 1995 Japanese survey data and fiscal capability is calculated using 1995 fiscal year budget figures and standard government-calculated measures of local fiscal need.

(i) **Model 1** Respects existing local boundaries (except those of government-designated large cities) and regional boundaries, so consolidation is undertaken without dividing up existing cities and municipalities. The aim here is to divide large city regions to create units that would offer citizens greater autonomy, for these city regions are often extremely large. For example, imagine a case in which there are 100 people in a city legislature. If the population is 1,000,000, each legislator represents 10,000 people. This is equivalent to a municipality of 10,000 people having only one elected official. This would clearly be a case of insufficient representation that would not be attractive from the perspective of popular participation. Nevertheless, the reason I raise the possibility of respecting existing regional boundaries is that such boundaries, and the regional cooperative associations within them that provide services such as garbage collection, fire prevention, and libraries, are held to be relatively meaningful. They have been the foundation of many attempts to create wider local blocks.

(ii) **Model 2** relaxes the first Model's no-partitioning rule, to allow the partition of municipalities close in size to government-designated large cities. It adds the condition that the population of the largest municipality must be no more than twice that of the smallest.

(iii) **Model 3** takes into account Kurokawa's (1997) recommendation of maintaining regional economies, by creating municipalities based on populations of approximately 300,000 people. With one exception, Model 3 does not allow for new municipalities to cut across existing metropolitan or prefectural lines, and, like Model 2, population differences must be kept within the 2:1 range.[2]

Simulation Results

Compared to the projections derived from Model 3, the consolidated municipalities in Models 1 and 2 are rather large, with average populations greater than 500,000; varying between 400,000 and 800,000 people per consolidated municipality. Disparities in fiscal strength in Models 1 and 2 are also not as large as in Model 3. However, creating these larger municipalities leads to wide disparities in the physical space consumed by municipalities. The Tokyo and Kansai metropolitan municipalities are so densely populated that they take up little space. But the Hokkaido municipality combines a large number of existing administrative units, like Okinawa and Nagasaki prefectures (which include many islands), resulting in a very large geographical space. In Model 3, which maintains existing economic blocks, there are about 300,000 people per municipality. Compared to Models 1 and 2, the areas are smaller and closer to the actual shape of economically linked regions. On a different note, disparities between consolidated municipalities' fiscal capability scores are greater than those for Models 1 and 2.

Models 1 and 2 yield about 250 regions and Model 3 about 450. It is worth noting that the former is close to the number of seats in the House of Councilors (the less powerful Upper House) and the latter is close to the number of seats in the House of Representatives. Perhaps more important, the projections of municipal fiscal capability (a ratio comparing revenue to projected standard fiscal demands) under these reapportioned systems are better than actual ones.[3] My primary thought in consolidating municipalities is to strengthen local fiscal bases, but there is a short-term problem as a result of municipal consolidation. Cutting the total number of municipalities by one-half does reduce the number of executives and legislative members. But, by doing nothing more than consolidation, there is the danger of creating the illusion that it is possible to enact all necessary reforms in one fell swoop. It is clear from the chain of events throughout the postwar period that consolidation is not a complete solution. It is true that under consolidation, the number of municipalities will fall by half, but personnel expenses may not be halved immediately. The current civil service system is an obstacle to this. But, at the same time, because of the short-term costs of newly adopted regulations, the cost-saving effectiveness of consolidation will develop gradually over the long run. Moreover, with consolidation, expenses based on items like the maintenance of old local government offices will limit the ability to cut costs immediately.

Prospects: The Need for System Reform

Local decentralization is a process that reshapes the role of a country's local governments and, by altering limits on municipalities, should help create a closer connection between the people and local administration. Compared to unitary systems, multiple-level systems place great restrictions on municipal power, but it is increasingly important that decentralization of authority to municipalities occurs. However, achieving decentralization is complicated by the nature of the relationship between various governments.

It is important to remember that even with shifts of authority to municipalities, prefectural governments are important components of the Japanese governmental system. As the case of government-designated large cities shows, transferring power to municipalities may stimulate the hollowing out of prefectures. Whether prefectures should be abolished or reduced in number has been debated, but government planning has proceeded under the assumption that a multi-level system will be maintained. At the same time, the size of cities and towns has continued to expand since the Meiji era. Discussion over reforming local government has centered on plans to strengthen municipal fiscal affairs and make local administration more efficient. Just as it has been a problem in the past, difficulties in reflecting citizen preferences, alongside issues of disparities between municipalities, may arise within merged jurisdictions.

After some fifty-odd years of debate surrounding decentralization, it seems that Japan must engage in serious rethinking about its municipalities. Improving local autonomy requires debate not only on reforms in local fiscal administration, but also on matters of scale. Within debate over consolidation, attention must be given to substantial internal reform and the self-strengthening of municipalities themselves.

Finally, I would like to conclude with some recommendations for reforming this system and creating a healthier governmental financial system in Japan.

Reform 1: The Subsidy System

When requests are made for expenditure, they ought to be accompanied by plans to raise the revenue necessary to fund these programs. Where programs do not make explicit from where revenue will come, they should not be funded. Since this might lead to inflated revenue estimates on the part of municipalities, these estimates should be incorporated into their standard fiscal revenue. When revenues fall short, allocation taxes should be reduced by that amount. The opposite should hold as well: when more revenue than expected is collected, the municipality can use this amount as it sees fit. In any case, the key component in all this is that by forcing municipalities to consider revenue sources and amounts in advance, it will force them to become better planners, better managers, and more responsible administrators.

Reform 2: The Allocation Tax

To date, allocation taxes have been used to cover the whole difference between the fiscal expenditure and revenue. My proposal is to cut the proportion of the difference covered by 5% each year. In this way, while municipalities currently use allocation taxes to cover the expenditure-revenue gaps in their entirety, in the next fiscal year the allocation tax would only cover 95% of differences, then only 90%, and so on. Such measures should be used until the proportion falls to 75–80%.

Reform 3: Introduce a Minimum Tax Rate

Currently, each municipality can decide on tax rates between a standard, 'base' level and a maximum level, but lowering the rate below the standard is difficult.

For this reason, it is difficult to reward citizens, even in municipalities with good management rates. Establishing a minimum tax rate that is two-thirds of the base level, and allowing municipalities the discretion to reduce the rate, would remove this impediment. By lowering taxes in this way, municipalities are provided a means by which to attract citizens and businesses.

Reform 4: Local Responsibility

Since local bond offerings currently exceed 100 million yen, in effect institutional investors are the only ones able to purchase them. This system should be discarded, and the central government should be barred from reimbursing interest. If the market is then allowed to determine interest rates, open managerial competition should result.

Reform 5: Promote Citizen Participation

Under such competition over administrative policies, there is the potential for improvements or possibly the worsening of citizens' lives. Whatever the result in the quality of services, the resulting competition over administrative practices will bear on citizens, so policy decisions should be entrusted to citizen referenda. There is a caveat to this point, which is that there is always the danger that great public participation can grow violent. For this reason, certain basic citizens' rights should be established and insulated from the scope of referendum power.

Reform 6: Create Standardized Municipalities through Consolidation

Given the administrative competition I mentioned above, new municipalities should be created through the merger of municipalities that cannot effectively manage their affairs. According to my calculations, if we turn the current 3,400 municipalities into four or five hundred, this should create municipalities that, for the most part, can function independently. Without such mergers, delegation of authority to municipalities will not progress, and fiscal crisis will prove unavoidable in the long term.

Municipalities that reach defined population levels and fiscal capability scores ought to be capable of exercising broader authority. The Ministry of Home Affairs ought to take the lead in encouraging the maturation of local government and the development of a fiscally sounder municipal system. If it can do this, the Ministry will truly be an important force in the 21st century.

NOTES

1 This said, subsidy levels have been reshuffled to make them more uniform across Japan. This has occurred in line with administrative reforms that have been adopted since the 1980s. The process of increasing uniformity of treatment has continued since the economic bubble burst and throughout the difficult economic times that have followed. Yet the issue of lobbying national government bureaucrats by extravagant wining and dining has moved to front and center in the news recently. In this regard, there is concern that closeness to the national government might play a part in corrupting municipalities.

2 In the consolidation under Model 1 dividing existing cities is not permitted, but in several regions municipalities do cut across prefectural or metropolitan borders. Similarly, in Model 2, the consolidation divided existing cities in four cases and crossed existing prefectural/metropolitan lines in seven cases.

3 In Model 1, the largest population sits in the Tokushima prefecture region (at 838,000) and the smallest is in Senboku, excluding Sakai city (at 303,000), which gives a ratio of largest to smallest of 2.76:1. In Model 2, the largest population is the Kohoku New Town region (794,000) and the smallest is Nishinomiya (399,000). As required, the disparity in size here is only 1.99:1. In Model 3, the largest population is the Fukuyama area (414,000) and the smallest is Higashi Izu and Minami Izu (208,000). Here the population disparity is just under 2:1.

REFERENCES

Amakawa, A. (1994). The Local Administration System. In: M. Nishio, & M. Muramatsu (Eds.), *Administrative Politics – Volume Two: System and Organization*, (121–57). Tokyo: Yuhikaku.

Cabinet Public Relations Office. (March 1965). *Public Opinion on Ideas about Political Autonomy*. Tokyo.

Cabinet Public Relations Office. (March 1967). *Public Opinion Data on City Branches and Advocacy Offices*. Tokyo.

Hayashi, M. (1980). Problems of Local Governmental Compartmentalization and Aggregation in Japan. In: M. Hayashi, & K. Sane (Eds.), *Consolidating Localities and Local Autonomy*, (43–63). Tokyo: Kokon Shoin.

Hayashi, Y. (1999). *Local Finance*. Tokyo: Yuhikaku.

Japan, Ministry of Home Affairs (1996). *National Register of Service Outsourcing Entities*. Toky: Gyosei.

Japan, Ministry of Home Affairs. (1997). *White Paper on Local Finance*. Tokyo: Ministry of Finance Printing Bureau.

Japan, Ministry of Home Affairs, Local Administration Bureau. (1953). *Data on the Promotion of Local Administrative Consolidation*. Tokyo.

Japan, Municipal Local Autonomy Study Group. (1995). *Q & A: A Handbook of Local Administrative Consolidation*. Toky: Gyosei.

Japan Local Autonomy Society. (Ed.). (1989). *Recollections and Views on Japanese Local Autonomy*. Tokyo: Keibundo.

Jinno, N,, & Kaneko, M. (1998). *Give Localities a Tax Base*. Tokyo: Toyo Keizai Shinposha.

Kobayashi, Y. (Ed.). (1998). *Empirical Analysis of Local Autonomy: Japan, the United States, and Korea in Comparative Perspective*. Tokyo: Keio University Press.

Kobayashi, Y., Niikawa, T., Sasaki, N., & Kuwabara, H. (1987). *Realities of Local Government: A Survey of Leading Policy Makers*. Tokyo: Gakuyo Shobo.

Kobayashi, Y., & Kawamura, K. (1997). Empirical Analysis of the Central Local Government Connection in Japan. *Legal Studies 70(9)*, 43–64.

Kurokawa, K. (1997). Consolidating Localities. In: T. Kaneya (Ed.), *Localities in an Age of Decentralization* Osaka: Research Group of Local Administration Osaka

Local Administration Investigative Committee. (1952). *Survey Data on Local Administration*. Tokyo.

Muramatsu, M. (1988). *Local Politics*. Tokyo: Tokyo University Press.

Oishi, T. (1980). The Distribution of Independent Localities and Population Magnitude. In: M. Hayashi, & K. Sane (Eds.), *Consolidating Localities and Local Autonomy*, (64–84). Tokyo: Kokon Shoin.

Omori, W., & Satoh, S. (1986). *Local Government in Japan*. Tokyo: Tokyo University Press.

Osugi, S. (1994.) Administrative Reform and Local Governmental Reform. In: M. Nishio, & M. Muramatsu (Eds.), *Administrative Politics – Volume Two: System and Organization*, (285–326). Tokyo: Yuhikaku.

Prewar Bureaucratic System Study Group (I. Hata (Ed.)). (1995). *The Prewar Japanese Bureaucratic System: Organization and Personnel*. Tokyo: Tokyo University Press.

Research Institute of Problems in Local Administration in Osaka (1995). *Restructuring and the Internationalization of Osaka*. Osaka: Toho Press.

Sakaguchi, K., Mizuyama, T., & Kotani, T. (1980). The Problem of Maintaining an Independent Township in the Local Administration System. In: M. Hayashi & K. Sane (Eds.), *Consolidating Localities and Local Autonomy*, (252–276). Tokyo: Kokon Shoin.

Sakata, T. (1998). *A Scenario for Decentralization*. Tokyo: Gyosei.

Takayose S. (1998). *A Fiscal Analysis of the New Local Autonomy*. Tokyo: Keiso Shobo.

Tanaka, K. (1996). *Administrative Reform: The Complete Works of Contemporary Administrative Law 10*. Tokyo: Gyosei.

Tsuji, S. (Ed.). (1976). *Administrative Politics – Volume Two: Administrative History*: Tokyo: Tokyo University Press.

Tsunematsu, S., Maeda, M., & Sasaki, N. (1994). Topics in the Evaluation of City Nuclei and Broad Local Regionalization: Readings on Local Administrative Law Reform Plans. *Local Fiscal Affairs 481*, 11–38.

APPENDIX A: THE DEBATE OVER DECENTRALIZATION IN JAPAN

In July 1995, the Local Decentralization Initiative Act was passed in Japan. Today the national government continues to progress in reforming the national-local division of labor. Under decentralization, municipalities are given greater independence, and authority over local revenue is transferred from the central government to them. At the same time, the conditions municipalities face are fairly severe. According to the 1996 fiscal year White Paper on local financing, local governments owed more than 100 trillion yen. The repayment of this money will inevitably have a tremendous influence on future politics in countless key ways. Decentralization without a serious effort to reform the financing system will be harmful to the Japanese people, the country's ultimate rulers.

When we look at how local government officials feel about the movement toward decentralization, the number of mayors, legislative leaders, and local government financial officers who thought that the national government should take the bulk of responsibility for local governmental development was a paltry 20% or so. The remaining 70% plus thought responsibility should be given to municipalities themselves. The largest group of respondents felt that for local fiscal affairs to develop, municipalities, and not the national government, must be at the heart of this development.

Yet, we must raise questions about the desirability of extremely fast decentralization. If implemented too quickly, decentralization could lead to problems, with concern over the potential for scandal at the local level accentuating the danger inherent in rapid change.

APPENDIX B: THE DEBATE OVER DECENTRALIZATION IN SOUTH KOREA

June 27, 1995 will go down in Korean history as the first time all four local-level elections were held simultaneously. Up until that point, even though local legislators were chosen in public elections, local executives were appointed by the national government. The old system did not provide 'fully' autonomous municipalities. This was vividly exemplified during the Park presidency, which was a hotbed of corruption, when the local governmental system was abolished. Given that local-level elections had not previously been held simultaneously, 1995 marked a great opportunity to begin a new national-local government relationship, as 'full' local autonomy in Korea was only just beginning. The survey used in South Korea aims to make clear the degree to which local government may or may not find success in their earliest forays into autonomous governmental control.

Like Japan, the problems that the South Korean local governments face indicate the need for boundaries to be established between the jurisdictions of the local and central governments. Nevertheless, unlike Japan where decentralization is premised on the experience (whether successful or not) of a local governmental system in the postwar period, in Korea the capability of local governments remains untested.

APPENDIX C: THE DEBATE OVER CENTRALIZATION IN THE U.S.

The main problem for municipalities in the USA is that there is a huge disparity in their levels of resource and administrative services. This leads some commentators to call for greater higher-tier governmental overview, with improved systems of resource allocation to assist municipalities with low resources and high expenditure demands. A symbol of the pressure resulting from inadequate resources on municipalities is the oft-cited fall in governmental spending on elementary education and social welfare. For example, the survey found that, in contrast to Japanese and South Korean mayors, a very large number of U.S. mayors want to cut their social welfare budgets below existing levels. With regard to health care and insurance, the inclination among Japanese and Korean mayors is typically to increase funding, while U.S. mayors merely seek to maintain current commitments. These differences are due partly to different tax systems. But most striking of all is the tremendous influence U.S. taxpayer lobbies are able to wield over the American system of administration.

APPENDIX D: THE SURVEYS OF LOCAL LEADERS

The data for this study are based on a mail survey conducted among Japanese, South Korean, and U.S. local governmental leaders. To summarize:

The Japanese Survey

The Japanese survey was conducted in all 664 cities in Japan in 1995. It focused on mayors, speakers of city legislatures and government section chiefs responsible for fiscal affairs. By the end of the year, 365 mayors (55% of those who had been sent the survey), 380 speakers (57%) and 380 financial section heads (64%) had responded.

The U.S. Survey

Because of the timing of the fiscal year, the U.S. survey began slightly later than the Japanese and Korean ones. The U.S. survey focused on a random sample of cities of 20,000 people or more, alongside ones surveyed continuously by Terry Clark. The surveys were sent to mayors, city legislators, and CAO/City Managers in the first two months of 1995. Ultimately, responses came from 174 mayors, 276 legislators, and 584 CAO/city managers.

The South Korean Survey

The South Korean survey was sent to all mayors, speakers of city legislatures, and local policy planners in South Korean cities and broader geographical ward regions (230 city wards). The sample here is made up of the 135 mayors, 90 speakers and 149 planners who had responded by the end of 1995.

11. DO NEW LEADERS RISK SHORTER POLITICAL LIVES? ASSESSING THE IMPACT OF THE NEW POLITICAL CULTURE

Terry Nichols Clark

Is political volatility rising? Many observers answer yes, political volatility is rising as part of the new politics. Examples that are cited to indicate this include the decline of traditional parties and programs (Lipset, 2000), the rise of new parties, and of new types of leaders inside older parties – like Tony Blair, Bill Clinton, and Gerhard Schroeder (Clark & Hoffmann-Martinot, 1998; Giddens, 2000), the weakened role of trade unions, and their separation from traditional left parties in many European countries. These changes may foster political volatility if voters grow less socially constrained by strong, overlapping, nested organizations, of the sort that the German Social Democratic Party launched in the nineteenth century, with unions, clubs, housing, job location, company stores, vacation villas, etc. linked to the Party. These organizational types flowered via Scandinavian social democratic parties, alongside the communist parties and governing regimes of East Europe and the former Soviet Union. The recent decline of such overlapping organizations should expose individual citizens to more 'cross-pressures', which Lipset (1981) has shown leads to attitudinal ambivalence and less consistent voting. That is, the decline of class politics and all-encompassing political parties should heighten political volatility. Note that this and related propositions are formulated as *ceteris paribus*, probabilistic statements. They assume all else is equal but recognize that these changes are only one part of a more complex process.

Volatility should similarly rise if clientelism declines. Patron-client exchange networks remain the classic form of political system in most of the world, especially in areas recently based on agriculture. Such clientelism has been most studied in locations like Ireland and southern Italy (e.g. Bax, 1976; Littlewood, 1981), and in U.S. cities with immigrants from such Catholic peasant regions (e.g. Clark, 1975). But clientelism came under attack globally with the spread of anti-clientelist concerns in the 1990s. Indeed from Taiwan to Naples to Chicago, eliminating 'political corruption' became a major issue. Practices ranging from giving favors to making campaign contributions, that were standard operating procedures even in 1990, led to jail sentences by 2000 in some locations like France and Italy. If traditional clientelist networks linking voters and patrons decline, social individualism should increase, which in turn should increase political volatility and turnover amongst elected officials.

A similar social constraint logic can extend to political leaders: if they rise through, and are nominated for public office by strong parties, they are likely to continue to support the party, and stay in office themselves longer. But a competing proposition emerges with new politics trends: leaders in weak party systems can adapt with more agility to changing citizen concerns, especially if these concerns are local and leaders can respond to them locally, without waiting for changes in national party programs. If changing citizen concerns increasingly drive the new politics, even a strong party cannot guarantee long tenure. Young and dissenting party members in moderately strong party systems like France may act independently of their national party program, but this is risky. Many New Mayors from the 1980s, like Alain Carignon and Michel Noir, ended their careers in jail. This situation suggests that if a party is too strong and ideologically rigid, its candidates may serve shorter terms, since they may adapt poorly to their changing constituents. Party strength may thus be curvilinear in its effects on tenure: positive as long as the social loyalty impact keeps increasing with party strength, but negative when it grows so strong that the leadership cannot adapt rapidly to new challenges. How fast other elements of society change, particularly voting preferences, should thus shift the impact of party strength on tenure in political office.

Does shorter tenure mean more citizen-responsive leaders? Some reformers advocate 'term limits' (prohibiting an elected official from serving more than one or two terms of office) and shorter terms of office (not the six years of traditional French mayors, but more like one or two years, as for some U.S. mayors), making the argument that leaders who serve more briefly are more responsive to their constituents. But note a possible logical problem with this argument. It is clarified if we contrast two situations. The first is one where leaders serve just one or two short terms, and then leave office. New leaders then replace them. In the second situation, the same officials serve for more years, and may be re-elected several times. The normative argument is that elections require citizen input and thus enhance citizen responsiveness. But an unintended consequence of a more frequent replacement of leaders is that over time it creates more years of 'lame duck' leadership; that is longer periods when leaders hold office but know that they will not continue after their term expires. Indeed it is classic for leaders to implement unpopular decisions in such years since they will not be exposed to a vote on their decisions. Eulau & Prewitt (1973) were thus concerned about the potential non-responsiveness of many council members they studied in the San Francisco Bay area who left office voluntarily after just one or two terms. These were typically local business people and professionals (especially lawyers) who returned to private practice after an interlude as elected politicians. Such quick turnover, they held, decreased competition among candidates for votes and thus decreased citizen responsiveness.

Much of the U.S. literature implicitly conceives of responsiveness as flowing from decisions of individual political leaders in a mainly local context, like the San Francisco area example. But if we look internationally, political contexts vary, and more diverse leadership arrangements affect local responsiveness. For instance, in most of Western Europe, national parties dominate local politics (see Figure 1.1; also Saiz & Geser, 1999). Local elections then in good part become local referenda on national issues. If the key issues for campaigns in voters' and candidates' minds are national, then local responsiveness suffers (Simonson & Robbins, 2000). Such a national logic may also have a strategic element: if citizens perceive that national leaders will be unchanged by their vote, they may vote against candidates of the same party as the national leaders to protest national policies. Alternatively, voters may abstain. Such concerns led a British observer (Williams, 1993, p. 107) to comment that ". . . the prospect of losing office is effective as a restraint on arbitrary government in Britain only in so far that the opposition is taken seriously as a potential government. Throughout the 1980s such a condition did not pertain." My British coeditor elaborated on these issues: [at the moment] "– people want to give Tony Blair a kick in the teeth – as they did with the devolution votes in Wales and Scotland – now I very much doubt that many of the same people will vote for anything except Labour in a national election – so local tenure, in countries where voters decide on such issues using national criteria, can produce strange results" (Keith Hoggart, email, April 24, 2000).

These interrelated concerns, and competing hypotheses, are pursued in this chapter, by examining the number of years that local officials hold office in countries where we have original survey data on local political system characteristics (the number of observations thus shift by variable in tables below).

CHANGING ROLES OF POLITICAL PARTIES

It is not clear that *local* parties are weakening. Indeed, many contributors to Saiz & Geser (1999) report that in much of Western Europe national parties have penetrated further down to local levels in the last decade or two. Nationalized local party activities grow with increased grants and staff support, newsletters, and similar activities to mobilize local voters; albeit often around national themes. In partial contradiction, Clark & Hoffmann-Martinot (1998, chs. 2–5) suggest that voters have grown more alienated from political parties, exacerbating low voter turnout. More citizens now declare themselves to be 'independent' of all parties, and traditional party programs drive away many votes – especially younger and more educated ones – who favor more emphasis on new issues like women, ecology, and human rights. The multiple processes that are changing parties are underway simultaneously in many countries, making 'party impacts' harder, although not impossible, to test. We can capture many but not all of these ideas using data from the Fiscal Austerity and Urban Innovation Project. Below we assess the relative impact on turnover of these two competing tendencies.

Tenure: Turnover, Term of Office, Years Served

There is a small literature that seeks to explain public official turnover, or years served in office. Much past work is from countries like Spain, France, and Italy, where in Napoleonic manner traditionally strong ministries of the interior maintained detailed demographic-type records on local officials (name, age, gender, party, occupation, and more). These data were long kept secret, but have gradually been released to social scientists (e.g. Bettin, 1993; Martinotti & Melis, 1990). These researchers have mined these data for national overviews of local leadership, being encouraged in this task by the absence of original survey data, especially in years before surveys were available. Recent U.S. surveys are considered below.

Modeling Political Tenure

Has tenure of local elected officials declined due to increased political volatility? To study this question properly, we need to detail some causes and correlates of what we may term 'tenure' or 'years served' by present incumbents. This is approximately the inverse of 'turnover', but as we compare large numbers of officials cross-sectionally here, rather than exploring holders of the same positions over time, we do not literally measure turnover. Tenure is thus a more felicitous label. To assess the sources and consequences of new political rules like volatility, shifting citizen preferences, or political individualization via weaker social contacts, we need: (1) to conceptualize and measure our key variables – tenure and new rules; (2) to specify a model which ideally would include all major factors affecting tenure; and (3) to determine if new rules decrease tenure.

Tenure is straightforward to conceptualize and measure. In this study it is the number of years served by the incumbent in the present office, as mayor or council member. Note however that the measure comes from incumbents, some of whom may remain in office for many years after answering our survey. Hence responses are downwardly biased indicators of years actually served. This is the same for most one-time surveys of leaders, including most of the literature cited below (e.g. DeSantis & Renner, 1994). Our data come from local government officials who were surveyed as part of the Fiscal Austerity and Urban Innovation (FAUI) Project. The FAUI surveys have covered more than 7,000 local governments in some 30 countries, but not all officials responded, and not all data have been pooled for comparable analysis. A more limited set of cities is thus used here. More detail on the FAUI Project is in Clark and Hoffmann-Martinot (1998, p. 168ff).

New Political Culture (NPC) is a concept to help interpret changes in current political systems, sometimes labeled the 'new politics' or, especially in the U.K., the 'Third Way' (Clark & Rempel, 1997; Giddens, 2000). A central, open question is whether such new rules bring more volatility and shorten political tenure. We started with party and social change hypotheses, suggesting strong parties increase tenure, but added that as citizens embrace more of the New Political Culture, and engage in more issue politics, socially or politically-connected leaders may be voted out earlier if they do not adapt to the NPC issues that are embraced by their constituents.

We examine several individual measures which capture elements of this new political culture, among both citizens and leaders, using FAUI surveys of mayors and council members as informants, plus census-type sources: education of the population, age of the mayor, and more. We also create a composite index of New Fiscal Populism (NFP) in five countries. This variable is first introduced in this paper, and incorporates several more discrete elements that should distinguish cities in terms of their political culture. Our five components are:

1. fiscal conservatism (a preference for lower spending) +
2. social liberalism (on abortion and sex education) +
3. citizen responsiveness +
4. group responsiveness −
5. an emphasis on public goods policies (rather than private goods which are more desirable for clientelist leaders) +

Various of these elements have been considered by many past analysts, but these five were first introduced as a distinct combination for defining four types of political culture in Clark & Ferguson (1983).[1] The four types were defined by distinct positive or negative contributions of these five underlying components (deep structures), as illustrated here by + and − signs for the New Fiscal Populist type. The other three types are classic left (e.g. European social democrat), classic right (U.S. Republican), and ethnic/clientelist politics (as in many agricultural areas, like southern Italy).

In conceptualizing the New Political Culture (Clark & Hoffmann-Martinot, 1998), fiscal conservatism was omitted as a defining characteristic, due to its lesser salience, especially in Western Europe in the 1970s and 1980s, in contrast to the USA, where it was central to the taxpayers' revolt of the 1970s (and was thus included in the definition of New Fiscal Populism in Clark & Ferguson, 1983). But by 2000, fiscal conservatism seems on the rise again among citizens in much of the world. This is perhaps spurred by globalization and the spread of 'neo-liberalism' and other ideas broadly analogous to the New Fiscal Populism that were identified in the USA in the 1970s. Fiscal conservatism is also the quintessential change of left parties, moving them from their classic views of strong national government toward the market and social individualism of the NPC. Hence we include fiscal conservatism here as a defining characteristic and label this our New Political Culture index or NPCIXA. We have computed NPCIXA and similar indexes for the three other types of political culture (traditional left and right, and clientelist/ethnic) from responses of each mayor to items necessary for computing these indexes, which are summarized in the Appendix.

RESULTS

Tenure varies considerably across nations (Table 11.1). French and Norwegian mayors serve over 13 years on average, while Australian mayors serve just 4.2 years (to be precise, these were the years served to date at the time of the survey). Yet mayors in some countries like the UK are quasi-ceremonial officials, since party 'leaders', with no formally recognized position on local councils, are in reality the main council decision-makers. We thus include all

Table 11.1: Number of Years Served by Mayors Varies Considerably Across Countries

	Number of Years in Office	Number of Officials Surveyed
United States	5.1	1030
Canada	7.2	100
France	13.5	176
Japan	8.3	681
Norway	13.4	457
Australia	4.2	241
Israel (mayors only)	11.6	19
Italy	5.6	53
Britain (council members)	6.9	593
Britain (party leaders)	9.1	27
Poland, 1988	6.5	243
Israel (mayor+council)	10.6	94
Total	9.3	3714

Source: International Fiscal Austerity and Urban Innovation Project Surveys, results only for countries that asked tenure questions. Most surveys conducted in the late 1980s or early 1990s.

British council members, who average 6.9 year incumbencies, as well as the 27 who reported that they were council leaders, and had 9.1 years of tenure. This is apparently one of the first international surveys of local leader incumbency, so simply presenting such descriptive results provides new knowledge. In the USA, the International City Management Association (ICMA) surveyed 914 mayors on their incumbency in 1991, finding almost the identical result, for 5.1 years of tenure was average using the FAUI data and 5.4 years was the ICMA figure (DeSantis & Renner, 1994, p. 39).

Sources of Variation

The few recent studies of local officials' tenure have found weak predictors. DeSantis & Renner (1994) analyzed 13 variables in their ICMA survey, but found only one significantly affected tenure – those localities that held local elections simultaneous with state and national elections reported lower tenure by mayors. A more limited U.S. survey of city manager tenure found two sorts of factors important. First, localities with more political conflict, and whose leaders reported a lack of confidence with the city manager, had shorter tenure managers. Second, managers who were more mobile in their aspirations and had a Master of Public Administration degree were more likely to move, especially from localities with more conflict (DeHoog & Whitaker, 1990, p. 370, p. 374). This second study covered just the state of Florida and explored managers rather than mayors or council members. Managers are far more likely to move than elected officials, but the conflict results might plausibly hold more generally and are broadly consistent with the political volatility proposition.

Numerous possible explanations of tenure were explored, but those that were modeled most carefully emerged as important sources of the New Political Culture in Clark & Hoffmann-Martinot (1998, ch. 4). These eighteen characteristics were refined through a series of analyses (moving from correlation through multiple regression). Nine characteristics significantly predicted mayoral tenure when all countries were pooled (Table 11.2). Tenure was greater in cities with high incomes compared to others in the same country (INCOME1), with Catholic mayors, more inequality in educational attainment by citizens, council members who reported that they were not responsive to citizens, and a higher percentage of blue collar workers. These are generally characteristics encouraging more traditional class and clientelist politics than the New Political Culture. By contrast, tenure was lower if the locality had more persons working in professional and technical jobs (that is, classic post-industrial occupations, whose incumbents tend to be more critical of politicians and supporters of the NPC). Similarly, tenure was lower in countries with higher incomes (INCOME2). Mayors with more years of education also had lower tenure, presumably since their educational qualifications gave them more job opportunities, as well as perhaps a more critical outlook. Cities with stronger parties had mayors with shorter tenure, which contradicts the hypothesis about a strong party providing more stable political leadership, but is broadly consistent with the competing NPC hypothesis that cities with stronger parties respond less well to their constituents, who thus vote out leaders more often, shortening their tenure.

Two regression models were computed using a larger and smaller number of explanatory variables. The larger model permits testing more variables, but since many individual variables are often moderately interrelated, distinguishing their relative impacts can demand closer analysis. Hence we completed the smaller model in Table 11.3. The main change is that the New Political Culture Index (NPCIXA) became nearly statistically significant in the smaller model, and it has a negative coefficient, indicating that mayors have shorter tenure in cities with higher New Political Culture scores. This is significant at just the 0.106 rather than the standard 0.10 probability level, i.e. the probability that the relationship could have occurred simply by chance.[2] Nine other plausible causes of tenure were found to be insignificant, such as population size. These are listed in Table 11.2.

Similar analyses were computed for each country separately. The results are weaker than the all-countries' model, due to more limited variation and fewer

Table 11.2: Sources of Variation of Mayoral Tenure: All Countries Combined, Large Model

Dependent Variable: Number of Years as Mayor						
Full Equation Results	R	R Square	Adjusted R Square	Std. Error of the Estimate		
	0.63	0.40	0.33	5.39		
	Independent Variables and Their Coefficients	Unstandardized Coefficients	Standardized Coefficients	t statistic	Significance Level	
	(Constant)	-119.28	29.11	-4.10	0.00	
	% persons in professional/technical	-0.83	0.18	-4.49	0.00	
	SPOIX2	-0.05	0.02	-2.28	0.02	
	INCOME2	0.00	0.00	-1.90	0.06	
	Education of Mayor (in Years)	-0.27	0.16	-1.73	0.08	
	local wealth=PC inc in USA	0.00	0.00	-1.34	0.18	
	POPULATION 1989	0.00	0.00	-0.87	0.39	
	Independents: combo noparty/missing/mayormissing	-0.86	1.35	-0.64	0.53	
	group activity index	-0.01	0.03	-0.59	0.56	
	Mayors average spending pref for all areas	0.00	0.03	0.08	0.94	
	NPCIXA	0.00	0.02	0.29	0.78	
	social conservatism index	0.02	0.02	1.02	0.31	
	Favorable Response to Citizens by Mayors	0.03	0.02	1.07	0.29	
	MEDIA USE: (IV140+IV141)/2	0.02	0.01	1.49	0.14	
	Atkinson - Education, alpha=4	0.13	0.08	1.75	0.08	
	% BLUE COLLAR	0.11	0.06	1.92	0.06	
	Favorable Response to Citizens by Council Members	0.04	0.02	2.02	0.05	
	catholic mayor=1 non-catholic mayor=0	4.96	1.32	3.75	0.00	
	INCOME1	2.74	0.56	4.89	0.00	
Note: These are results from multiple regression of data from the international Fiscal Austerity and Urban Innovation Project. See the list of variables for sources and definitions						

cases. Yet several country-specific results did emerge. For instance, stronger parties (the same SPOIX item as in Figure 1.1) increased the tenure of mayors in Japan, but not in other countries. This makes sense if we consider that Japan has less NPC than other countries in general. It also has relatively stable local policies, more limited demands from organized groups and citizens, and general stability (Clark & Kobayshi, 2000). Consequently strong local leadership by the Liberal Democratic Party helps local officials stay in office longer in Japan, which is consistent with the isolation from turbulence hypotheses stated above. Localities with more highly-educated populations had mayors with shorter tenure in the USA, but with longer tenure in Japan and Norway. Education of the mayor consistently led to shorter tenure except in Italy, where more educated mayors

Table 11.3: Sources of Variation of Mayoral Tenure:
All Countries Combined, Simplified Model

Dependent Variable: Number of Years as Mayor

Full Equation Results	R	R Square	Adjusted R Square	Std. Error of the Estimate
	0.36	0.13	0.12	6.20

Independent Variables and Their Coefficients	Unstandardized Coefficients	Standardized Coefficients	t statistic	Significance Level
	B	Beta		
(Constant)	16.87	1.82	9.28	0.00
Education of Mayor	-0.63	0.10	-6.66	0.00
NPCIXA	-0.01	0.01	-1.62	0.11
Atkinson - Education	-0.05	0.04	-1.39	0.17
Independents	-0.74	0.80	-0.93	0.36
MEDIA USE	0.02	0.01	1.74	0.08
Catholic mayor=1	2.14	0.75	2.86	0.00

Note: This is a model with a smaller number of independent varibles which shows a significant effect of the NPCIXA Index. Its rise to significance from the larger model in Table 11.2 is probably due suppression of the Index by several other variables moderately related to it. Results are from multiple regression of data from the international Fiscal Austerity and Urban Innovation Project. See the list of variables for sources and definitions served longer. This Italian result is probably explained by the more closed system of political leader selection in the past (the survey was from the late 1980s), which included training in party schools and numerous other seniority requirements by parties. These helped form what Italians commonly termed a 'political class' (Recchi, 1997). Of those Italians who chose to pursue political careers, the more highly educated remained in office longer, demonstrating that the costs of entry into politics were so high that if this path was chosen incumbents stayed in politics. Moreover, if political entrants trained longer, they held tenure longer, and shifted less often to non-political occupations than in other countries. This illustrates the closed character of the political system before it was reformed in the early 1990s in efforts to open it to non-professionals.

We created three measures of income for cross-national comparison. INCOME1 was the locality's percentile score compared to all other localities within the same country, calculated in the relevant national currency. Affluent localities compared to others in Japan had officials with longer tenure. Again this is consistent with the general conservatism of local government in Japan compared to other countries, with this pattern upheld in the more affluent Japanese localities. Our second measure was the national-level income per capita, in U.S. dollars; the score for which was the same for all cities in a country. Generally, cities in nations with higher incomes had local officials with shorter periods of tenure (we used this in part to permit inclusion of the last income measure). The third income measure was local income per capita, which was converted into U.S. dollars. It had no impact on tenure, except in Japan where it was positive.

Seeking to separate political party and leader effects, we pursued the idea that an entrepreneurial mayor may deviate from the national party program. We thus created measures of the policy distance of each mayor by taking the absolute value of the difference between her or his responses to fiscal and social policy items from those of all other mayors in the same party in the same country (PRFAVG and SOCCONS). We next divided by that mayor's reported party strength measure (SPOIX). Running these policy difference items in correlations and regressions, we found limited impacts on tenure. Of the 10 countries for which we had full information, just three showed modest results: in Norway and Australia, mayors who deviated further from the national average did increase their tenure, supporting the NPC policy entrepreneurship hypothesis. But for Israel the result was the opposite: if mayors took more distant policy stands, their tenure declined, which supports the view that a strong national party punishes deviant local officials. However, this result held for just 19 Israeli mayors who answered all of the survey items. There were no significant statistical relations when the combined total of 94 Israeli mayors and council members was examined. The scatter plot for Norway is included here as there are more

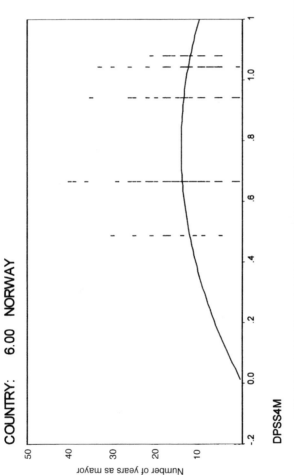

Figure 11.1: Scatterplot of Mayor's Policy Distance from National Party.

cases than the other countries (Figure 11.1). The Norwegian results suggest a slight curvilinear pattern as posited above: mayors that deviate moderately from the national party position may increase their tenure, but are punished if they go too far. As this is for just one country, we should be cautious about generalizing, but it illustrates a possible approach to elaborate.

It would be intriguing to pursue such ideas by combining further analysis of this sort with some ethnographic work on selected cases like those of the French New Mayors or council members attempting to deviate from their national parties. In exploratory conversations on these points with British council members in the early era of New Labour under Tony Blair, the battles between old and new Labour tendencies were most explicit, as was the fear of junior council members to deviate from the local party whip. They reported that they would destroy their careers and not be slated again if they expressed significant dissent. This may be too low a level at which to expect much deviance, however. The party leader in a locality, or a group of officials in many towns acting as a sort of caucus, would be more likely to take on the national leadership. Yet at least one outstanding case of a deviant individual exits. This is that of Ken Livingstone, who openly battled his national party leadership, was not slated, ran independently, and was elected Mayor of London in 2000.

DISCUSSION

Most results for the individual factors affecting tenure are broadly consistent with the New Political Culture hypotheses and with results in Clark & Hoffmann-Martinot (1998: ch 2 and 4). Mayors enjoy longer tenure in locations with more blue-collar workers and less educated mayors, and shorter tenure where the local workforce is more engaged in professional and technical occupations. Tenure is shorter for mayors who score higher on the New Fiscal Populism Index (that is, reporting that they are more fiscally conservative, socially progressive, responsive to citizens, less responsive to groups, and stress public goods). While the overall power of these results is fair to good by most social science standards, clearly here, as in most of life, we confront relatively modest relations, that are less than deterministic patterns. Put differently, we by no means find the strong and consistent result which many observers fearfully posited: that political conflict and volatility have so increased that leaders have much shorter public lives. The simplest reason this pattern remains weak may be that ambitious leaders who wish to remain in office have learned to adapt to the NPC context. Hence, mayors who report that they use the media more actively stay in office longer; a factor more important in the USA than other countries.

Some well-publicized extreme cases may clarify more general patterns. For instance several of the most nationally-visible New Mayors elected in France in the 1980s ended in jail or were legally removed from office, like Michel Noir from Lyon and Alain Carignon from Grenoble (Balme et al. 1986–87). Both rose to high national government positions. They also actively sought to remake their national parties and challenged traditional leaders. They ran into many roadblocks. The dramatic controversy surrounding the impeachment of President Bill Clinton by the U.S. Congress may have grown out of similar concerns. That is, highly innovative and politically ambitious persons often lead very risky careers (Clark, 1999). Slighted persons around them easily grow jealous and mobilize against the innovators.

How important are all the factors we could measure using these standard indicators? In the larger Table 11.2 model, they explain about 33% of the variance in mayoral tenure (the adjusted R^2). This tells us conversely that 67% is not explained. Hence, even if a mayor had the bad fortune to be in a location where every one of the 18 characteristics worked against her or him, s/he could still overcome their effects. Other things not included in the table, like a forceful personality and creative choice of policies, are still roughly twice as important as the 18 standard items. This is good news for public officials of all policy persuasions.

NOTES

1 Measures of these five in the Appendix are PRFAVG, SOCCONS, Group Activity Index, CITRESP, and Break Rules.

2 The other changes from Table 11.2 to Table 11.3 are that the Atkinson education index falls below, and Media use by the major rises above the significance level. Mayors who use the media more actively enjoy longer tenure.

REFERENCES

Balme, R., Becquart-Leclercq, J., Clark, T. N., Hoffmann-Martinot, V., & Nevers, J-Y. (1986–87). New Mayors. *The Tocqueville Review 8*, 263–278.

Bax, M. (1976). *Harpstrings and Confessions: Machine-style Politics in the Irish Republic*. Assen: Van Gorcum.

Bettin, G. (Ed.). (1993). *Classe Politica et Citta*. Padova: CEDAM.

Clark, T. N. (1999). The Clinton Paradox: A WordWide Perspective. *Frontier* April/May, 17.

Clark, T. N., & Ferguson, L. C. (1983). *City Money: Political Processes, Fiscal Strain and Retrenchment*. New York: Columbia University Press.

Clark, T. N., & Rempel, M. (Eds.). (1997). *Citizen Politics in Post-Industrial Societies*. Boulder: Westview.

Clark, T. N., & Kobayshi, Y. (Eds.). (2000). *The New Political Culture in Japan*. Tokyo: Keio University Press.
Clark, T. N., & Hoffmann-Martinot, V. (1998). *The New Political Culture*. Boulder: Westview.
Clark, T. N. (1975). The Irish Ethic and the Spirit of Patronage. *Ethnicity 2*, 305–359.
De Hoog, R. H., & Whittaker, G. (1990). City Manager Turnover. *Journal of Urban Affairs 12*, 361–377.
De Santis, V., & Renner, T. (1994). Term Limits and Turnover Among Local Officials. In: *The Municipal Yearbook*, (36–42). Washington, D.C.: International City Management Association.
Eulau, H., & Prewitt, K. (1973). *Labyrinths of Democracy*. Indianapolis: Bobbs-Merrill.
Giddens, A. (2000). *The Third Way and Its Critics*. London: Blackwell.
Littlewood, P. (1981). Patrons or Bigshots? Paternalism, Patronage and Clientelist Welfare in Southern Italy. *Sociologia Ruralis 21*, 1–17.
Lipset, S. M. (1981). *Political Man: The Social Bases of Politics,* 2nd edition. Baltimore: Johns Hopkins University Press
Lipset, S. M. (2000). The Decline of Class Ideologies: The End of Political Exceptionalism?. In: T. N. Clark, & S. M. Lipset (Eds.). *The Breakdown of Class Politics: A Debate on Post-Industrial Stratification*. Baltimore: Johns Hopkins University Press.
Martinotti, G., & Melis, A. (1990). Gli Amministratori Communali (1973–1987). *Amministrare 28*, 2–3.
Recchi, E. (1997). *Giovani Politici*. Padova: CEDAM.
Saiz, M., & Geser, H. (1999). *Local Parties in Political and Organizational Perspective*. Boulder: Westview.
Simonsen, W., & Robbins, M.D. (2000). *Citizen Participation in Resource Allocation*. Boulder: Westview.
Williams, T. (1993). Local Government Role-Reversal in the New Contract Culture. In: R. J. Bennett (Eds.), *Local Government in the New Europe*, (95–108). London: Belhaven.

Appendix: Variables in Model, Definitions and Sources

Tenure: Years as Mayor = the Dependent Variable in the Regressions	MAYYRS=Years served as mayor
INCOME1, 2	**INCOME1 INCOME2** Income data came from each national census in national currencies. To analyze differences within each country, we created a standard (Z) score for each city that is its deviation from the country mean: Income1. The Income 1 scores were also used in some pooled crossnational analyses to permit comparing cities that are rich in national terms. But to capture crossnational differences in income, we used Income2, which is the national average income per capita in 1986 from the World Bank. Both income measures are included in some analyses that then capture both types of income effects: relative national and absolute cross
% persons in	Professional and technical workers as percent of the labor force taps a postindustrial economic dimension, IOCPROF
Catholic mayor	CATHMY=Dummy variable for Catholic mayors (1=Catholic, 0=not)
Atkinson - Education, alpha=4	Ir = Atkinson's Index of Inequality. The logic is the same for income or education categories, but is simpler to explain with income. Yi = income of income class i (e.g., persons from $5,000 to $10,000 annual income) Y = mean income for the social unit (the city) Pi = proportion of income earned by income class i =alpha = the exponential coefficient that the analyst can vary to specify the rate at which inequality affects the overall Index
	The index subtracts from one the ratio of a city where each individual is equal in income to the mean of the actual income distribution in that city. Thus Ir declines as the income distribution grows more equal. The magnitude of this decline in Ir is also affected by alpha. As alpha rises, more weight is attached to transfers at the lower end of the distribution; at infinity, only transfers at the bottom are taken into account. Thus one can vary the alpha coefficient to emphasize, or deemphasize, the effect of a small improvement in income by the poorest income group. Although Atkinson and most interpreters discuss this as the researcher's choice, one can assign alpha values to match different political cultures—a leftist political culture is presumably more sensitive to the lowest income groups and should value a higher alpha more than would conservative political cultures. Accordingly, four alpha values were chosen in computing each of the Atkinson indexes from .5 to 0., -.5, and –1, a range of alpha values recommended by Atkinson and researchers who have refined the index.
Education of Mayor (in Years)	MAYED=Years of education for the mayor, calculated in similar manner to IMEANEDU above.
% BLUE COLLAR	IPCTBLUE=Percent of the local labor force who are blue-collar workers, i.e. who work as manual laborers or "production workers".
group activity index	How active are several local organized groups, GRPACT. But note that this includes all kinds of organized groups like business and unions, so it is not a clear measure of NPC.
local wealth=PC inc in USA	IPCINCT=Mean level of income per capita for the city's population. In countries where this was not directly available, such as France, the best available measure of local wealth was used.
MEDIA USE: (IV140+IV141)/2	How important the media are in elections and general news coverage, MEDIA
Favorable Response to Citizens by Mayors/Councilmembers	The mayors and councilors were asked: "Sometimes elected officials believe that they should take policy positions which are unpopular with the majority of their constituents. About how often would you estimate that you took a position against the dominant opinion of your constituents?" Responses: 0=never or almost never 25=only rarely 5-=about once a month 75=more than once a month 100=regularly.

Do New Leaders Risk Shorter Political Lives? 283

Social conservatism index	SOCCONS summed two social liberalism/conservatism items: Q19 (V131) Would you be for or against sex education in public schools? (Circle one number) 1 For 2 Against 3 Don't Know/Not Applicable Q20 (V132) Do you think abortion should be legal under any circumstance, legal only under certain circumstances, or never legal under any circumstances? (Circle one number) 1 Under any circumstances 2 Under certain circumstances 3 Never legal 4 Don't know/Not Applicable. Soccons6 differs from Soccons; it adjusts for missing data in France
POPULATION 1989	LIPOPT=Log of city population for a year near the time of the survey, often 1985 or 1990
Independents	Respondents who are not political party members
Mayors average spending pref for all areas	PRFAVG - Average spending preference of mayor on thirteen FAUI items. The key item was Mayor Q4: "Please indicate your own preferences about spending. Circle one of the six answers for each of the 13 policy areas. 1 Spend a lot less on services provided by the city 2 Spend somewhat less 3 Spend the same as is now spent 4 Spend somewhat more 5 Spend a lot more 6 DKr Don't know/not applicable. Policy areas: All areas of city government, Primary and secondary education, Social welfare, Streets and parking, Mass transit, Public health and hospitals, Parks and recreation, Low income housing, Police protection, Fire protection, Capital stock (e.g., roads, sewers, etc.), Number of municipal employees, Salaries of municipal employees. The mean for a city was calculated if the mayor provided answers for a minimum of four policy areas, since some countries and mayors omitted items.
SPOIX2	Strong Party Organization index, based on items from the mayor indicating how often meets with local party officials, how active they are in local campaign activities, and how important they are in slating local candidates.
NPCIXA	New Political Culture Index, sums five components: npcixa = - soccon6a - prfavga + citrespa + iv130a - grpres where: citrespa = mean(iv101, - iv142). Citizen Responsiveness, citrespa sums two items. IV101=Q9 Please indicate how often the city government responded favorably to the spending preferences of the participant in the last three years. The city has responded favorably to individual citizens 1 Almost never 2 Less than half the time 3 About half the time 4 More than half the time 5 Almost all the time 6 DK Don't know/not applicable V142=Q30 Sometimes elected officials believe that they should take policy positions which are unpopular with the majority of their constituents. About how often would you estimate that you took a position against the dominant opinion of your constituents? 1 Never or almost never 2 Only rarely 3 About once a month 4 More than once a month 5 Regularly

Mayor's Spending Preferences, prfavga = mean(iv7 to iv19).
Where IV7 to 19 correspond to the spending areas listed above for prfavg.

iv131a = .4 * iv131.
iv132a = .6 * iv132.

Social Conservatism, soccon6a = mean(iv131a, iv132a), where the two items are those used in the Soccons6 index above.

grpres sums iv90, iv91, iv93, iv94, iv95, iv96, iv97, iv99, iv100.

Group Responsiveness, Grpres indicates the city responds to multiple organized groups (rather than citizens as a whole). Q9 Please indicate how often the city government responded favorably to the spending preferences of the participant in the last three years. (Circle one of the six answers for each of the types of participants.)
The city has responded favorably
1 1   Almost never
2 2   Less than half the time
3 3   About half the time
4 4   More than half the time
5 5   Almost all the time
6   6   DK Don□t know/not applicable
1 PUBLIC EMPLOYEES AND THEIR
UNIONS OR ASSOCIATIONS v90
2 ORGANIZATIONS CONCERNED WITH
LOW-INCOME GROUPS AND FAMILIES v91
3 HOMEOWNERS? GROUPS OR ORGANIZATIONS   v92
4 NEIGHBORHOOD GROUPS OR ORGANIZATIONS   v93
5 CIVIC GROUPS (e.g., THE LEAGUE OF WOMEN VOTERS)   V94
6 ORGANIZATIONS CONCERNED WITH MINORITY GROUPS   V95
7 TAXPAYERS' ASSOCIATIONS   v96
8 BUSINESSES AND BUSINESS-ORIENTED GROUPS OR ORGANIZATIONS
(e.g., CHAMBER OF COMMERCE)   V97
9 THE ELDERLY   v99
10 CHURCHES AND RELIGIOUS GROUPS   v100

Public Goods Emphasis in Policy: iv130a = .4 * iv130.

V130=Q18 □If a political leader helps people who need it, it doesn't matter that some of the rules are broken." What is your feeling? (Circle one number) 1 AGREE 2 DISAGREE 3 DON'T KNOW

Note: These variables were developed from international FAUI files. More detail about the FAUI study and most variables is in Clark and Hoffmann-Martinot, 1998: ch 4.

LIST OF CONTRIBUTORS

Rachel Ashworth
Rachel Ashworth is a Research Associate working within the Public Services Research Unit at Cardiff Business School, Cardiff University. Her previous research on local government was on party competition and spatial equity whilst her doctoral research focused on mandate theory at the local level. She has published in journals such as *Public Administration, Policy and Politics* and *Local Government Studies*. Her current research interests include investigating political and organizational accountability.

George Boyne
George Boyne is Professor of Public Sector Management and a Distinguished Senior Research Fellow in the Public Services Research Unit at Cardiff Business School, Cardiff University. He has published widely on local government structure, finance and performance in journals such as *Public Administration, Public Administration Review, Public Choice, Urban Affairs Review* and *Urban Studies*. He is the author of *Constraints, Choices and Public Policies* (JAI Press, 1996), *Public Choice Theory and Local Government* (MacMillan, 1998) and editor of *Managing Local Services: From CCT to Best Value* (Frank Cass, 1999). His main current research interest is the explanation and evaluation of organizational performance in the public sector.

Terry Nichols Clark
Terry Nichols Clark is Professor of Sociology at the University of Chicago. He holds MA and Ph.D. degrees from Columbia University, and has taught at Columbia, Harvard, Yale, the Sorbonne, University of Florence, and UCLA. His last books are *The New Political Culture* and *Urban Innovation*. He is Coordinator of the Fiscal Austerity and Urban Innovation Project.

Bas Denters
Bas Denters (1954) is associate professor in political science at the Faculty of Public Administration and Public Policy at the University of Twente in Enschede (The Netherlands). His current research interests are urban politics and urban policy, local democracy and local and regional governance. In 1988 he received

the Annual Award of the Dutch Political Science Association (NKWP) for the best political science publication in Dutch political science in 1987 for his Ph.D. thesis. He has published in *Acta Politica, the European Journal of Political Research, Quality & Quantity, Public Administration* and *Policy Studies Review*.

Keith Hoggart
Keith Hoggart is Professor of Geography at King's College London. He has also taught at the University of Maryland and Temple University, as well as holding a visiting research post at the University of California, Berkeley. His primary research interests are on local policy implementation, housing provision (especially social housing) and links between migration and social change. Included amongst the books he has published are *People, Power and Place: Perspectives on Anglo-American Politics* (Routledge, 1991) and *Rural Europe: Identity and Change* (Arnold, 1995).

Yoshiaki Kobayashi
Yoshiaki Kobayashi is Professor of Political Science at Keio University. In addition to receiving his Ph.D. and teaching at Keio University, he has also been a visiting professor at the University of Michigan and the University of Cambridge, as well as a visiting researcher at Princeton University and the University of California, Berkeley. His main research interests are on local autonomy, elections and voting behavior public choice, both in Japan and from a comparative national perspective. Among his major publications are *Modern Japanese Politics 1955–1993, Modern Japanese Elections, Public Choice*, and *Quantitative Political Science*.

Volker Kunz
Volker Kunz is Professor of Political Science at the University of Mainz. His recent research has included a major study of municipal policy outputs across Germany, which received an award from the German Institute for *Urbanistik* in Berlin (the Carl Goerdeler-Prize for Research on Local Government and Politics). His research interests include central-local relations, local government organization, decision theory and comparative economic analyses. At present he is engaged in two major research projects, both concerning the determinants and consequences of social capital in western democracies.

Per Arnt Pettersen
Per Arnt Pettersen is Professor of Political Science at the Norwegian University of Science and Technology in Trondheim, Norway. He has also held positions in both Oslo and Bodø, Norway, and has been visiting professor at a variety

of universities in the United States. His research and publication concerns issues of political participation, political attitudes and perceptions, and studies regarding the development of the Norwegian welfare state.

Lawrence E. Rose
Lawrence E. Rose is Professor of Political Science at the University of Oslo. He has also held positions at the Norwegian Institute for Urban and Regional Research, as well as the University of Virginia and Stanford University in the United States. His research interests and writing focus on issues of political participation and contemporary democratic citizenship, local government and politics, and comparative public policy.

Jefferey Sellers
Jefferey M. Sellers is Assistant Professor of Political Science at the University of Southern California. He has taught at Yale University, the Humboldt University in Berlin and Boston University. In a series of journal articles, and a forthcoming book, entitled *Governing From Below: Urban Regions and the Global Economy* (Cambridge University Press), his recent research analyzes the political, economic and social sources of policymaking within cities throughout Europe and the United States.

Carlos Nuñes Silva
Carlos Nuñes Silva is Professor Auxiliar in Geography at the University of Lisbon. His primary area of teaching and research focus on local government politics and policy determination. He has published on urban governance, local government policies, local public finance, social housing and urban planning history. His publications include the 1994 book *Política Urbana em Lisboa 1926–1974* (Livros Horizonte, Lisbon).

Melanie Walter
Melanie Walter is Assistant Professor of Political Science at the University of Stuttgart. Her main research interests focus on political attitudes and political behavior in western democracies, particularly in Germany, on empirical evaluation of the theoretical propositions of democracy (such as responsiveness and elite research) and on political methodology. Her most recent publication in the year 2000 is: Die Deutschen Politiker in der Sicht der Bevölkerung – Wert-, Macht- oder Funktionselite? In: *Wirklich ein Volk?: Die Politischen Orientierungen von Ost- und Westdeutschen im Vergleich*, edited by J. W. Falter, O. W. Gabriel and H. Rattinger. Opladen: Leske and Budrich.

Research in Urban Policy

Edited by Terry Nichols Clark
University of Chicago

Volume 4, Politics of Policy Innovation in Chicago
1992, 232 pp. $73.25/£47.00
ISBN 1-55938-057-8

Edited by **Kenneth K. Wong**, *University of Chicago*

CONTENTS: Introduction: Policy Innovation in the Political and Fiscal Context. Part I. Innovation in Taxation and Services. Taxes in Chicago and Its Suburbs, *Terry N. Clark, Daniel K. Crane, Ann L. Kelley, Joanne Malinowski, Mellissa Pappas, and Gregory Wass.* Privatization in Chicago's City Government, *Rowan Miranda.* Part II. Linking Social Equity to Development. The New Politics of Sports Policy, *John Pelissero, Beth Henschen, and Edward I. Sidlow.* Decentralization of Policymaking Under Mayor Harold Washington, *Robert Mier, Wim Wiewel, and Lauri Alpern.* The Politics of Housing Policy, *Barbara Ferman and William Grimshaw.* Part III. Varieties in Governance Reform. The Implementation of Health Care Planning, *Grace Budrys.* Local School Reform: The Changing Shape of Educational Politics in Chicago, *James G. Cibulka.* Choice in Public Schools: Their Institutional Functions and Distributive Consequences, *Kenneth K. Wong.* Part IV. Political Constraint and Policy Choices. Strategies for Confronting Austerity, *Terry N. Clark.* Coping with the New Fiscal Federalism: Changing Patterns of Federal and State Aid to Chicago, *Lawrence B. Joseph.*

An Imprint of
Elsevier Science

www.jaipress.com

Research in Urban Policy

Edited by Terry Nichols Clark
University of Chicago

Volume 5, Local Administration in the Policy Process:
An International Perspective
1994, 290 pp. $73.25/£47.00
ISBN 1-55938-361-5

Edited by **Carmel Coyle**, *University College Dublin*

CONTENTS: Introduction, *Carmel Coyle*. Organizational Fragmentation in Greek Local Government: Municipal Bureaucracies and Municipal Enterprises, *Paraskevy Kaler-Christofilopoulou*. Irish Local Administration in the National and European Policy Process, *Carmel Coyle*. Sub-National Bureaucracy in the United Kingdom: The Scottish Office, *Richard Parry*. The Role of the British Local Government Chief Executive: A Response to Challenge, *Alan Norton*. Toward a Centralized Municipal Labor Market and Back? Local Employees Unions and the Labor Market in Finland and Scandinavia, *Voitto Helander*. How Central are Decentral Personnel Politics? Growth and Bargaining in Denmark, *Finn Bruun*. Communal Administrators: The Swedish Case, *P.O. Norell*. Reform Government and Fiscal Austerity Strategies in American Cities, *Lynn M. Appleton*. City Workers and Fiscal Cutbacks: Cross-National Comparisons, *Lynn M. Appleton and Vincent Hoffmann-Martinot*. Explaining the Privatization Decisions Among Local Governments in the United States, *Rowan A. Miranda*. Administrative Culture: A Mode of Understanding Public Administration Across Cultures, *Ishtiaq Jamil*.

An Imprint of
Elsevier Science

www.jaipress.com

Research in Urban Policy

Edited by Terry Nichols Clark
University of Chicago

Volume 6, Constraints, Choices and Public Policies
1996, 256 pp. $73.25/£47.00
ISBN 1-55938-896-X

By **George A. Boyne,** *University of Wales, Cardiff*

A large number of studies have compared policy outputs across national and local political systems. This book evaluates the validity of four theories of public policy making. Two of these theories suggest that governmental decisions are the product of external constraints and are determined by political choices which reflect either the ideology of the ruling political party or the self interests of politicians and bureaucrats.

The book develops a framework for evaluating the validity of these theories and applies it t evidence drawn from analyses of local policy variation in the UK. Little of the evidence can be taken at face value because of theoretical and methodological flaws in the empirical studies. Nevertheless, by subjecting the evidence to critical scrutiny it is possible to draw conclusions on both the empirical validity of the theories and their logical coherence.

An Imprint of
Elsevier Science

www.jaipress.com

Research in Urban Policy

Edited by Terry Nichols Clark
University of Chicago

Volume 7, Solving Urban Problems in Urban Areas Characterized by Fragmentation and Divisiveness
1998, 312 pp. $73.25/£47.00
ISBN 0-7623-0464-2

Edited by **Fred W. Becker** and **Milan J. Dluhy**,
School of Policy and Management, College of Urban and Public Affairs, Florida International University

CONTENTS: Solving Problems in Urban Areas Characterized by Fragmentation and Divisiveness: An Overview, *Fred W. Becker and Milan J. Dluhy.* Part I: Socioeconomic context of the metropolitan region. Part II: Fragmentation, divisiveness, and governmental organization. Part III. Fragmentation, divisiveness, and law enforcement. Part IV: Fragmentation, divisiveness, and health and social services. Part V. fragmentation, divisiveness, infrastructure and regional development.

An Imprint of
Elsevier Science

www.jaipress.com